Scientific Writing and Publishing

Knowing how to prepare, write and publish high-quality research papers can be challenging for scientists at all stages of their career. This manual guides readers through successfully framing and presenting research findings, as well as the processes involved in publishing in learned journals. It draws on the author's wealth of practical experience, from working in academic research for over 40 years and teaching scientific writing in over 20 countries, to gaining insights as a journal editor. Well-written and logical, it provides clear step-by-step instructions to enable readers to become more effective at writing articles, and navigating difficulties related to journal submission, the review process, editing and publication. It comprehensively covers themes such as publication ethics, along with current topics including Open Access publishing and preprint servers. This is a useful, user-friendly guide for graduate students, early career scientists and more experienced researchers, particularly in the life and medical sciences.

Denys N. Wheatley is a retired academic who has spent over 50 years researching the cellular and molecular basis of cancer. He has published more than 300 papers in learned journals, and more than 100 editorials, commentaries and essays, often on issues involved in scientific writing, editing and publishing. This manual is his seventh book. He has sat on many editorial boards.

His first involvement with the publishing of learned articles came through *Biological Abstracts* in the 1960s, before computers were on the scene. In the mid 1990s, Denys became editor-in-chief of *Cell Biology International (CBI)*, until retiring in 2016. In the late 1990s, he was heavily involved in moving towards electronic (online) publishing. After forging a strong link with BioMed Central in London, working as an independent editor, Denys founded *Cancer Cell International* (2002) and *Theoretical Biology and Medical Modelling* (2006). In 2005, he was asked to take over editorship of *Oncology News*.

Shortly after taking over *CBI*, Denys set up BioMedES UK (www .biomedes.biz) because a 'level playing field' was needed to give non-native English-speaking authors a better chance of getting their papers accepted by ensuring that they were presented in the best possible idiomatic English within the correct scientific context. The manual addresses all that this entails, providing information and training that universities and other organizations fail to deliver. Having taught courses in ~20 countries, he knows why this manual is important, not just in the writing of first-class papers, but also in the complex processes on the publishing side that authors should know in much greater depth.

Scientific Writing and Publishing

A Comprehensive Manual for Authors

Denys N. Wheatley
BioMedES Ltd

CAMBRIDGE
UNIVERSITY PRESS

CAMBRIDGE
UNIVERSITY PRESS

University Printing House, Cambridge CB2 8BS, United Kingdom

One Liberty Plaza, 20th Floor, New York, NY 10006, USA

477 Williamstown Road, Port Melbourne, VIC 3207, Australia

314–321, 3rd Floor, Plot 3, Splendor Forum, Jasola District Centre, New Delhi – 110025, India

103 Penang Road, #05–06/07, Visioncrest Commercial, Singapore 238467

Cambridge University Press is part of the University of Cambridge.

It furthers the University's mission by disseminating knowledge in the pursuit of education, learning, and research at the highest international levels of excellence.

www.cambridge.org
Information on this title: www.cambridge.org/9781108835206
DOI: 10.1017/9781108891899

© Denys Wheatley 2021

First published 2021

Printed in the United Kingdom by TJ Books Limited, Padstow Cornwall

A catalogue record for this publication is available from the British Library.

ISBN 978-1-108-83520-6 Hardback
ISBN 978-1-108-79980-5 Paperback

Cambridge University Press has no responsibility for the persistence or accuracy of URLs for external or third-party internet websites referred to in this publication and does not guarantee that any content on such websites is, or will remain, accurate or appropriate.

Contents

Acknowledgements

After spending many years editing scientific and medical papers, having founded three journals, and developed a biomedical editing company (BioMedES – www.biomedes.biz), I have now spent 3 years or more putting together a manual on the subject of writing and publishing a scientific paper, in particular the primary research paper. Most novices and young authors nowadays seem to have to acquire the considerable skills involved by osmosis, and there is no question that these topics are seldom touched upon in graduate or early postgraduate programmes. This negligence in training of researchers is unforgivable, and is the reason why this manual has been written. It has not been hastily put together, but written, rewritten and honed many times to be as comprehensive as possible without being excessively tedious in its detail. The chapters present a logical sequence as to the way in which a scientific paper is constructed, submitted and published. It is equally surprising that so many new researchers have little idea as to what happens during the publishing process, knowledge of which can always smooth the path of a paper towards acceptance. I hope readers will forgive me for using the same epigraph for almost every chapter, which states why learning properly how to write and publish papers is so important; this cannot be overemphasized. Indeed the career and status of almost all of us depend on our ability to communicate effectively and efficiently. Science is all about communication. It is an art to write a paper, but in science accuracy and integrity in communication are paramount. Training must be followed by practice, more practice and yet more practice.

My task has been aided by many colleagues, but first and foremost I wish to acknowledge the help at all stages of my assistant, Mrs Angela Panther. I want to thank Claire Hamilton and Shona Owen, who recently joined me at BioMedES, in getting the manual into excellent shape ready for publication. I thank Drs Paul Agutter and Mark Henderson for their very sound advice on, and insight into, many matters. The emergence of online publishing around the turn of the millennium meant that I had to seek assistance from other colleagues, who at that time were much more computer literate. This was a time of almost complete change in what, in practical terms, publishing a paper was going to be like henceforth. I particularly wish to thank Peter Newmark (director of BioMed Central at its inception) and later Delphine Grynszpan, also at BMC. Nevertheless, how one *writes and composes* a learned article remains the same as before digital publishing came in. My thanks also go to Kurt Albertine, Joan Marsh, Cynthia Jensen, Ed Hull and many others who have at times pointed me in the right direction. Finally, I want to thank my partner, Jean Fletcher, for her unswerving support and tolerance throughout.

Introduction

Writing and Publishing a Scientific Paper
An Overview

Opening Remarks

This book is designed to help you write, improve, submit, revise and publish high quality research publications or reports. Papers are the final products of your research, possibly relating to experiments or investigations that might have taken several years to complete. Published papers are crucial for helping you to become an expert in your chosen field, communicate to the world your important contributions to the advancement of knowledge, get the necessary feedback to improve your research plans and establish your career. Therefore we cannot emphasize enough the importance of learning to write your articles effectively. Remarkably few colleges, universities, companies and other organizations worldwide take the trouble to instruct their students, research staff and others in paper writing. They seem to expect people to assimilate the requisite skills from their mentors (who themselves often need training). This manual will instruct you how to go about writing a research paper. To keep it within reasonable bounds, it does not include much information on how to write in good idiomatic English (see A.10). There is a wealth of books on this matter, and a quick search online will lead you to many. This is true not only regarding non-native English speakers, as many native speakers often need help, particularly important in developing a personal style, so necessary if papers are to be less stereotyped in the presentation.

Part A of this Introduction is an overview or outline of what editors and publishers of learned journals will demand of you (yes, *demand*) if you wish to publish a significant article. The aim here is to provide enough guidance about paper writing to obviate the need for recurrent consultation of the chapters that follow. They will give the detailed advice for inexperienced writers, and are especially useful when the time comes to

3

revise and improve your draft manuscript before submitting a polished article; but the 'quick guide' presented here should help you to begin with the right mind-set.

(Note, however, that the sequence in which one should write a paper will be found to be different in Chapters 1–8 from the more familiar layout of a finished paper covered in this overview.)

Parts B and C of this Introduction will outline *Submission* and *Publishing* procedures, respectively. They present the most up-to-date details on how to take a draft paper through the complicated modern processes instituted by editors and publishers, which grow ever more sophisticated in terms of software as publications in nearly all major journals are now being posted on the Internet. Once again, few people have an intimate knowledge of what goes on after having just submitted a paper and the procedures that lead to a final decision on acceptance or rejection of a paper. This also goes for the publishing procedures; although similar in many journals, the exact requirements of publishers can vary considerably. A good knowledge of what is going on after acceptance of a paper is therefore as important as many of the previous stages. If you do not go deeper into this manual than this Introduction, you will nevertheless have recognized and to some extent assimilated the essence of what is required of you in preparing a paper and how it thereafter proceeds to publication.

Part A   Major Issues in Drafting a Research Paper

A.1   Research: Having the Right Mind-Set

There are two challenges that anyone embarking on research will face. One is the daunting task of *writing up* a research project for submission to a learned journal. This manual will help to make it less daunting by guiding you through the process. While it might be especially aimed at the relative novice, many experienced researchers did not receive adequate training on this process. As emphasized in the early chapters, a primary research article is the final product of experimentation, possibly the outcome of months, often years, of diligent work. The main reason for having an explicit manual on scientific writing was summed up in an article (Wheatley, 2018) published after a conference in Philadelphia on the topic of preparing scientific papers for publication, which addresses the problem of a declining standard in presentation.[1] The associated papers are very helpful.

The other challenge preceded the one above, which was taking on a research project in the first place. Some novices might have been fortunate enough to have had an induction on how to prepare mentally and practically in order to meet the challenge of carrying out experimental work. The less fortunate entered research and then found out how to go ahead by osmosis – picking up the necessary skills from others as you go along. It is important to know how best to approach research, and recognize the characteristics that should be developed in order to enter on a research project with confidence. Clearly every project will have its own requirements, but there are *basic skills* common to all projects.

What are some of these fundamental characteristics that can be developed? Key attributes include the following:

(1) Having an inquisitive mind that wants to know more about the universe and, in particular, our own world from both its physical and biological aspects.
(2) Seeing and seizing the opportunity of gaining greater insight into a phenomenon worth pursuing.
(3) Having a good memory and powers of observation.
(4) Having the ability to organize thoughts in a rational and logical manner.
(5) Remaining critical and sceptical of existing explanations of a phenomenon.
(6) Acquiring knowledge from what has been taught or published, along with discussions with your peers and mentors, remembering point (5).

Some people make 'natural' researchers, but everyone can acquire the necessary skills if properly motivated. The verb to *motivate* is one of the most important; many need help getting properly motivated, and this goes hand-in-hand with *inspiration* and *encouragement*. Thus for those wishing to progress, the correct mind-set is paramount – a good research project can be identified and approached with greater alacrity. A musician cannot give a good recital unless fully motivated to learn good technique, and then have the artistry to give exciting performances. When you finally succeed

---

[1] Wheatley D. (2018). Writing scientific and medical papers clearly. The Anatomical Record, **301**, 1493–1496.

in some research work, you get a similar buzz; writing about it then becomes an exciting challenge. The Nobel laureate Albert Szent-Györgyi once said of research that it is important 'to see what everyone sees, but think what nobody else thinks'; to have a mind of your own (see points (3) and (4) above). The starting point of a good paper lies in formulating a new hypothesis.

A.2   Getting the Right Framework

- Readers want to know, in the fewest possible words, what your research findings or studies are and why you wish to share them. You must have good and original information to communicate: a **message** to impart to – literally – the whole world.
- At the outset you must say what you hope to add to the sum total of human knowledge: a new method, a highly salient new finding or new data that call a hypothesis into question.
- Who are your intended readers? Make sure you have a clear answer to this. You can then write so that those readers will be interested in what you have to say and will understand it, regardless of whether they are few or many, specialists or laypersons. Think of a journal that would be appropriate for your message; this will help you write your paper. But leave the final decision about where to publish until later.

A.3   Introducing Your Topic

- The **Introduction** sets the scene for your paper, putting your work in the right context. Its aim is to explain why you did this piece of work. It must be relevant and focused, *not* a comprehensive review of knowledge to date.
- Imagine you are delivering a work report (seminar) to a group of people, for example in your own department, who have interests allied to yours and want to hear what you have to say. This will improve both your flow and your style. Start by setting out the background to the work very quickly, with no detail unless necessary (the Discussion is for details).
- The explanation for your choice of topic should lead you to formulate a hypothesis. State this briefly. Your readers may well know why you made this choice and are often experts in the same topic. But only put the essentials down, again without explanation unless absolutely necessary.

- At the end of the Introduction, some writers add a sentence about what they have discovered. This is usually unnecessary because your Abstract has done it already (see Section A.8).

A.4   Telling Others What You Did

- As succinctly as possible, describe the *experimental procedures* you used to find evidence for (or against) the hypothesis. This is the **Materials and Methods** part of the paper. (Some journals put this section after the Discussion, often in smaller font.) Most readers who are interested in your Methods section will want to know how your results were obtained and perhaps wish to repeat your experiments (for confirmation, and to see whether a consensus emerges: scientific and medical advances depend on consensus). Others may wish to check to see whether your experimental approach was a valid way of obtaining your findings.
- Usually, groups (experimental and controls) are compared. Seldom do biological findings reveal themselves in precise and unequivocal differences between such groups. You usually have to run many tests in duplicate or greater multiples. This means that statistics are required, so you need to tell your readers what statistical tests were applied and why.

A.5   Presenting the Results

- Give your **Results** in as orderly a fashion as possible. The aim is to marshal the evidence for your conclusions with maximum clarity. Do your data support or refute the hypothesis you are considering? The answer will enable you to display the information to best advantage.
- Collect all the relevant tables and figures (preferably as files of a PowerPoint presentation). This simplifies the task of finding the most suitable order in which to present them – as if you were giving that seminar.
- The logic of the presentation should be obvious to the reader. Keep the sequence in which you performed the experiments in mind as you write, but the logic might demand a quite different order – indeed, your final experiment may have been the most important and will be used as your leading point.

- Do not discuss points in the Results section unless absolutely necessary. The Discussion is where you will argue your case.
- If you have conflicting data – experimental results that are both for and against your hypothesis – you have to accept them at face value (nature does not lie). Present them as you found them, not as you wished or expected to find them. You can interpret them in the Discussion section.
- A suggestion based on experience: start with the Results section when you begin to draft your paper.

A.6   Discussing the Findings

- When you discuss your results along with other workers' findings, remember to maintain balance. It is best to select four or five major points for the **Discussion** rather than address every detail of the results. The lesser issues will fall into place if the main points are presented and argued logically and cogently.
- Some points you make will corroborate received wisdom, some will extend it with new knowledge and others will conflict with previous ideas, perhaps providing the next received wisdom. When conflict is found, or you have results that do not fit comfortably with a hypothesis, say so. At some later stage, you or another person might resolve the issue (and perhaps open up new ideas and avenues of research).
- Speculation should always be well-based and kept to the point. It should remain within the confines of how far your data, along with data from other people, allow you to go. Most editors delete wild speculation, so avoid it.
- There is an important reason for writing the draft of a paper, which is that you have had to put your work into a wider context throughout the discussion. This is the stage at which you sometimes, and quite suddenly, realize that your research missed something or some part of it was poorly designed. The exercise of writing up draws your attention to matters that could be improved and clarified, even if it means going back almost to square one (e.g. having to rephrase your hypothesis and/or perhaps do more experimental work before publishing)!

A.7   Coming to the End of the Writing of the Text

- **Conclusions** are generally the last one or two sentences of the Discussion – at most – telling what you have found. Some people prefer to separate the Conclusions from the Discussion, but this is not usually necessary.

- The **Acknowledgements** section is to thank your funding agency (state the source and grant number) and all those who helped you in one way or another, including anyone who critically reviewed your draft manuscript.

- Cite the **References** you have used, keeping to the most appropriate – comprehensive lists are better kept for reviews. Reference managing software is now in common use; use it to your best advantage and to accord with the style of the journal to which you will send your paper.

A.8   Preparing an Abstract

- It might seem odd to leave the **Abstract** until after the Conclusions, but you will be clearer about to what to say when you have completed the rest of the draft. By then, your thoughts on the content of the paper will have been thoroughly rehearsed and refined.

- Certain things are required in the Abstract, but you need not add much background (many writers present too much – to readers who know as much as they do). The Abstract must be short and to the point. There often is a limit to the number of words permitted (typically 150–200), which forces you to be succinct.

- Some journals require a *structured* abstract, set out as Background, Methods, Results and Conclusions. In such cases, the word limit is usually higher.

- The first sentence can provide context, saying something about the area of research dealt with in your paper. Methods do not usually have to be described; there is little room for detail, so make only a short general statement unless some new technique or unusual application is involved. The sentence that assumes most importance is the main finding and it must have real impact. Shorter sentences should follow, indicating how the finding was corroborated by evidence. It is not a good plan to put data into the abstract unless they are needed to make the issue clear (e.g. a striking difference between certain measurements). Your Results section will give the full details, so do not give away your best evidence in the Abstract. Like a newspaper hoarding, the Abstract should draw in the potential reader.

- Do not include references (citations) in the Abstract unless absolutely necessary (many journals forbid them altogether).
- The last sentence needs to indicate what your research has shown, i.e. a brief conclusion, the main message to be imparted.
- Immediately below the Abstract is the place where *keywords* and *abbreviations* should be placed.

A.9   Giving the Paper a Title

- Deciding the **Title** is the last job in drafting an article. It must include keywords that will guide literature searches relating to your field of interest to your paper.
- The Title also has to interest and engage the reader, again like a newspaper headline. Therefore, do not make it too long, and ensure that it does not give the game away by stating the final result (known as a *declarative* or *pre-emptive title*, otherwise readers will see it, note its simple message and may not bother to read any further!).

Note: Although the order of preparation – Abstract and Title coming last – may seem odd, you will find in later chapters that some parts of a paper are better drafted in a different order from the way they finally appear in a publication, as has just been done here. The reasons will be explained more fully in the relevant chapters.

A.10   Writing in Good English

- Use the simplest form of sentence construction, but be mindful of sentence length, avoiding a 'staccato' delivery of successive very short sentences.
- Keep your subordinate clauses to a minimum and preferably after the main verb (this is preferred in English, though some other languages and cultures differ).
- Most reporting involves the past tense; if you switch to the present, do so only when essential. Results *were* found (past), but these *lead* to conclusions (now, the present).
- Vary the way in which sentences start; avoid repetition. The use of 'We' (We did this, and we found that...) is permissible nowadays, but it is

irritating if long successions of sentences begin with the same construction. (In past times, personal pronouns (I and We) were not permitted.) After a number of *active* sentences it is a relief to use the occasional *passive*. For example, 'We put five tubes in the incubator...' can be changed to the passive 'Five tubes were put into the incubator...'. However, this can be difficult for non-native English speakers and is another skill that must be learned

It is not easy to express yourself with style in English. This book does not intend to offer lessons in the art of English usage per se. However, suitable articles and books are given in the Further Reading section at the end of the manual from a vast array of books on the subject that are listed on the web in many search engines.

Part B Final Preparation of a Paper before Submission

B.1   Revising and Redrafting

- Read your draft paper again, and improve it by removing superfluous words and phrases.
- Make sure all your co-authors have read and helped you revise the new draft.
- If you have a chance to present your final draft in the form of a seminar, do so. Close colleagues who are not co-authors are often valuable critics. Revise your paper in the light of their comments. You may be surprised by how much they make you want to change what you have written.
- Go over the rewritten draft to see whether any words can be changed to make the meaning more precise. When you reread this draft, make sure the paper as a whole – not just each phrase or sentence – conveys the message(s) you want.
- Get an independent colleague (or two) who is/are quite expert in the same topic to read your penultimate draft. See whether the suggestions can improve the paper. If they do, make the appropriate alterations.

B.2   The Final Stages before Submission

- When you are satisfied you have your ultimate draft, reread the **Instructions to Authors** of the journal to which you have decided to submit the article. Try to comply with every instruction. This will put you in favour with the referees and editors, and if the paper is deemed acceptable it will make it quicker and easier to proceed.

- Make sure you have the **full consent** of all your co-authors and the head of department or institute (if required). Then submit the paper to just one journal of your choice, usually accompanying it with a short letter stating that all authors have consented, drawing attention to any 'conflict of interests', and explaining anything the editor might need to know (for example, whether it has been rejected by another journal; in that event you might need to say why it was rejected, e.g. unsuitable choice, wrong subject area). You might also want to indicate why this particular journal seems the most suitable.

B.3   A Note on Ethics before Submission

Ethics will be covered more fully in Part C of this Introduction and Outline, but before submission you must accord with good ethical practice in presenting a scientific paper for publication. This means that you must know what is not permissible (ignorance is no defence):

- Most importantly, you cannot send a paper to several journals at one time.
- You must not submit a paper that substantively resembles an already published article, yours or anyone else's; in brief, you article must be original. It cannot contain pieces from already published work (including your own) without permission (copyright).

There is much more to be said about publication ethics (see Chapter 16), but these are the two crucial points.

Part C Getting Published

C.1   What Is Required

Many authors are unaware of how the publishing business works, or of the desiderata, e.g. the author's and editor's contributions to making the process quicker, simpler and more efficient by adhering strictly to the publisher's requirements. Being familiar with the practices of referees, editors and publishers can be very valuable for smoothing the path of an article to publication. These topics will be covered in detail from Chapter 13 onwards; here, the significance of each will be outlined.

- **Online, open access or conventional (hard copy):** there are different modes of publication, with modern practice moving inexorably towards the online option, although many journals also publish hard copy. Soon virtually all will be open access online.
- **Costs and waivers:** you need to know who will pay and what for. The alternatives are (i) the author (or his/her funding system) pays upfront on acceptance of a paper for publication; (ii) the reader pays – the individual, but more usually the institute, has to subscribe in one way or another.
- **Instructions to authors:** guidelines to authors from journals and publishers are highly prescriptive; compliance is the best policy, right down to punctuation, units, reference styles and abbreviations. The better ones provide their checklist, which may differ from your own; be alert to any differences.
- **Checks for duplication, plagiarism, ethics, etc.:** there are ethical and moral issues in scientific publishing. Ever-improving software identifies unethical practices, and it is important to be aware of what can and cannot be done.
- **Publication management:** it is helpful to learn the roles of people in editorial offices and publishing houses. Modern submission systems usually allow authors to track the latest version of a paper from submission to the final publication step.
- **Permissions, copyright, corrigenda, etc.:** there are problems with reusing other people's words and data, correcting mistakes as a paper enters print, withdrawing or retracting claims, etc. These will be considered in detail in Chapter 15. There is also the issue of ownership of the published paper (who retains copyright).

C.2   What Is Not Permitted

- **Fabrication** of data is a cardinal sin. It can lead to summary rejection of any future work by the author(s). In some flagrant cases it has led to criminal proceedings and imprisonment.

- **Plagiarism** is the theft from previous papers – an infringement of copyright – of text, figures, tables, indeed any part of other peoples' intellectual property seen as published articles. It also refers to ideas stolen from others. It is highly unethical and should be avoided at all cost. There is in fact no excuse for it because permission can usually be granted by authors and publishing houses who hold the copyright, as long as acknowledgement is made and references given to any part of an article you might wish to include in your own paper. This is free of charge. Modern editing software is sophisticated enough these days to detect plagiarism.
- **Removal of data** that fail to conform to a pattern apparent in tables or figures is unethical. This also applies to omission of data that ought to be included. This practice is an act of deception and can also be self-defeating: 'rogue' points in a graph can sometimes be explained and even uncover an important but unsuspected aspect of the problem. It is possible to detect manipulation, e.g. 'airbrushing' of images; such practices should be avoided.
- **Multiple submission** is totally unacceptable. Editors can easily check for it these days. Submit to only one journal at a time.
- **Duplicate publication** is also a serious error and can be detected very quickly with modern software. Very minor modifications to a paper by the same (group of) authors, without change of substance, is usually treated as duplication.

Before going on to an example of a paper (Chapter 2) and details of the way in which to assemble a paper (Chapter 3), there is one essential matter to discuss – the **hypothesis** – for you must have a crystal clear hypothesis on which all you have to write hinges. As you begin to write your paper, it is worthwhile making a note at the very top of the first page of the hypothesis under test. Stating it succinctly is an exercise in itself, but well worth the effort, even if you need to modify its wording as you proceed. More detailed advice can be found in Appendix 5.1 of Chapter 5.

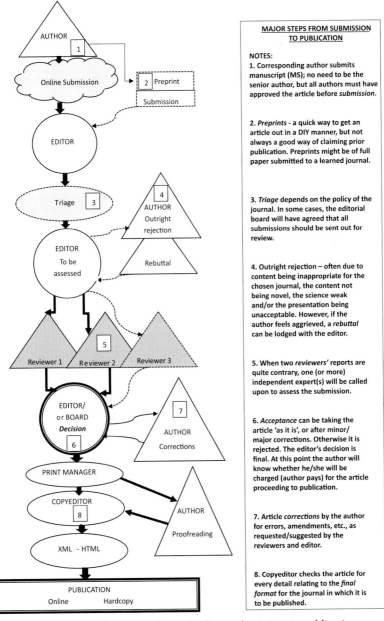

NOTES:

1. Corresponding author submits manuscript (MS); no need to be the senior author, but all authors must have approved the article before *submission*.

2. *Preprints* - a quick way to get an article out in a DIY manner, but not always a good way of claiming prior publication. Preprints might be of full paper submitted to a learned journal.

3. *Triage* depends on the policy of the journal. In some cases, the editorial board will have agreed that all submissions should be sent out for review.

4. Outright rejection – often due to content being inappropriate for the chosen journal, the content not being novel, the science weak and/or the presentation being unacceptable. However, if the author feels aggrieved, a *rebuttal* can be lodged with the editor.

5. When two *reviewers'* reports are quite contrary, one (or more) independent expert(s) will be called upon to assess the submission.

6. *Acceptance* can be taking the article 'as it is', or after minor/ major corrections. Otherwise it is rejected. The editor's decision is final. At this point the author will know whether he/she will be charged (author pays) for the article proceeding to publication.

7. Article *corrections* by the author for errors, amendments, etc., as requested/suggested by the reviewers and editor.

8. Copyeditor checks the article for every detail relating to the *final format* for the journal in which it is to be published.

Simplified flow chart illustrating the paths from submission to publication.

1 General Features of a Scientific Paper
Structure and Format

A research article is the *final product* of an investigation. It is a report that tells the world what you have done and found. On this account, it must be presented in the best possible way so that its message is to the point, clear and succinct.

1.1 Introduction: Why We Publish

The publication of your findings as a paper in a respected journal is the product of your research. It may seem a chore, but scientific endeavour is wasted if it is not presented to the world in the best possible manner.

All scientists want to present evidence for or against new ideas (hypotheses). There can be several motives:

- **Altruistic**. You see your work as contributing to the sum total of knowledge and a better understanding of nature. As scientists, we all want to construct reliable theories that stand the test of time: to be able to describe how a cell divides, what makes the economy tick, or how a drug works.
- **Personal**. In the interests of your personal reputation and your future in science, you want to claim a discovery. Publications meet the needs for recognition amongst peers, career advancement and the pursuit of higher degrees.
- **Research funding**. A good publication record is also crucial for attracting funding for future research – your own, and perhaps your colleagues' and your institution's. This is a major motive for scientific publication.
- **Commercial**. Your results might lend themselves to exploitation, in which case you need to establish intellectual property rights and perhaps secure a patent.

We would all like to draft our papers quickly and accurately, and then revise them effectively so they can be published and made available to the widest possible readership. However, many scientists dread writing. The purpose of this book is to help dissipate the dread and help you learn the necessary skills. Discovering how to organize the writing process will quickly improve your confidence.

In this first chapter we look at the structure of a typical paper. Later chapters will deal with the finer details of each component.

Nota bene: Some information, instructions and advice in this manual will be seen as repetitive. I am aware that this is not normally good practice, but it is *not unintentional*. Readers of a manual looking for guidance delve into different chapters/sections for specific information at different times. A manual is consulted, not read from cover to cover. Repetition in these circumstances will not be so self-evident; indeed, it can help to emphasize and reinforce many different points during the learning process.

Getting started

Have the right mind-set before attempting to write a first draft.

- People want to be told what you have found in the fewest possible words.
- You need to tell them at the start what you have added to the sum of knowledge.
- You need to decide the readership, i.e. which journal is likely to be most appropriate. This will ensure that you deliver your message in the right context for your readers.

1.2   The Framework of a Research Paper

The first question is always:

What message do you have for your readers? What new idea or information do you wish to share?

Originality is essential. Your message has to be placed up front, usually in the Abstract, without the need for much background. The message should be simple, short and straightforward, without conditional phrases. Avoid 'ifs' and 'buts' here; they can be considered later in the paper. Try to avoid in any place writing that you have 'demonstrated for the first time...'; if your paper has nothing original (new/novel) to tell, you should not be submitting it for publication. Should you submit it with the above phrase being written in the text, the editor should remove it as his or her job is to assure that papers provide new information – it is implicit.

Neither the Abstract nor the Title of the paper is the first thing to be drafted – see the end of this chapter (Section 1.9).

You will normally be expected to write the following sections: *Introduction, Materials and Methods, Results* and *Discussion*. Then you reiterate the main *Conclusions*. Appended to the draft will be *Acknowledgements, References, Figures* and *Tables*, and sometimes *Supplementary Material*. More and more journals now require you to include *Conflicts of Interest* statements and *Contributions* of the different authors.

These are the standard components of a typical research paper. The format has stood the test of time, though it has drawbacks and could no doubt be improved. (Some journals place certain sections in a different order, but the foregoing remains most common.) However, to get the best results, this is not normally the order you should adopt while you are writing the paper.

As we will see in later chapters, once you have decided on your key message, it is usually best to start with the Results section. However, at this stage, we are concerned only with the general outline of a research paper and the overall character of each section.

Below the Abstract there are usually two lists: a set of *Keywords* that indicate the subject matter and the field of the research; and the set of *Abbreviations* used in the paper. Keywords help readers to search databases and journal archives for related publications; the list is usually restricted to five to six items. The Abbreviations listed should not include commonly known items such as DNA.

A historical note about succinctness

Until well into the twentieth century, papers were printed on heavy mechanical presses. A compositor assembled the text letter by letter on little lead blocks, which were packed together by hand in an arrangement (back to front) that ensured they would produce the correct text when they were inked and pressed. For the compositor, succinctness was a virtue.

If the old method were still in use, only a fraction of the millions of papers now published each year would be printed. The electronic revolution and the Internet have changed the situation; words can be typed in quickly and directly by the author. However, succinctness should still be the rule, though for other reasons: editors and publishers do not want to face the laborious task of preparing long papers for publication, and you should expect the reader to be busy and have limited time. Readers do not want to wade through masses of verbiage – and they won't. Unless your paper is succinct it runs the risk of not being published, and even if it is published it might not be read. See this chapter's Appendix 1.1 regarding Word Reduction.

1.3   Introducing Your Paper

The main text of the paper begins with a brief Introduction. This sets the paper's aims in the appropriate context and prepares the reader to grasp the significance of the novel findings you will present. It should focus narrowly on the subject. (Sometimes it is difficult to identify the 'novelty', but we will deal with that in a later chapter.) If you have insufficient new data to communicate it is better not to proceed, except perhaps as a Preliminary Communication. This should state a hypothesis about the topic you are investigating and explain why your evidence to date supports that hypothesis (or otherwise).

A lengthy Introduction dilutes the message and reduces the impact on the reader. Beware of including anything that will be repeated in the Discussion (see below). Include only the most relevant background in the Introduction.

Remember who your reader is

Here is another reason why your Introduction should be brief and circumspect: most of your readers will already know at least as much about previous work in your chosen field as you do. A full review of the history (background) would be pointless and tedious for those readers. They don't want to be told what they already know.

1.4   How the Hypothesis Is Tested: Materials and Methods

For most people this is the most tedious part of the paper, but it is indispensable. Readers may want to compare and/or repeat what you have done and to develop the findings in their own way. Corroboration and refutation are the means by which science progresses; they depend on a precise grasp of the methods of investigation that have been used.

Many technical procedures are now routine. When possible, cite previous papers detailing accepted methods; you need not provide every last detail, but you must specify any modifications you have used. Accuracy is essential, so check the details – a wrong unit or a decimal point out of place can create problems. The SI system for units should be universal, and many units have accepted abbreviations (e.g. cm for centimetre, h for hour). These should be used – but not included in your Abbreviations list, which is for less commonly known items. Unless your paper is purely descriptive the Results section will be rich in numerical data, so the Materials and Methods must include the statistical procedures you have applied. Materials and Methods sections have a characteristic structure that has become almost universal; for example, the 'Materials' subsection is written first and 'Statistics' last.

How much detail do you need?

The least that can be said in this section the better. All the essential information must be there, but full details of techniques now in regular use are not required. For example, antigen–antibody reactions are so commonly used that one would not dream of explaining the chemistry of antigen–antibody binding in a Materials and Methods section, or indeed anywhere in the paper.

1.5   Results: the Pivot of the Paper

Readers want to see the new data you are offering to support or refute your hypothesis. The Results section ought not to be embellished with discourses on each finding, so include as little discussion here as possible.

Almost everything requiring further explanation can safely be left to the next section, where the findings will be discussed in themselves and in the light of previous publications. The clearer and punchier the Results section, the happier your reader will be. It is sometimes useful to summarize the essence of your findings in one or two sentences at the end of the section to prepare readers for the Discussion, especially when you have presented numerous data, but no more is required at this stage.

Presenting the data

Data relevant to sustaining or refuting your hypothesis need to be presented as straightforwardly as possible. Remember, some data will not fit comfortably with your idea, and may not yet be explicable. There may even have been some controversy or negative evidence. It is best to include such information, an issue we will deal with in a later chapter.

1.6   Discussing the Findings

It is pointless to reiterate the results when you open the Discussion, especially if you have summarized them at the end of the previous section. Unfortunately, this habit is becoming more widespread. The purpose of the Discussion is to tell the reader whether your data are likely to prove or disprove your (the) hypothesis, to present the relevant arguments, and to consider them in relation to previous publications. Indicate where caution in interpretation is needed, and qualify your conclusions as necessary. Few findings ever become solid fact – 'set in stone' – so there will always be 'ifs' and 'buts'. A paper is a state-of-the-art communication, a small step in the progress of knowledge.

A long-winded Discussion, occupying pages of text with little or no structure, should be avoided. A lengthy Discussion that lacks an internal logical sequence of arguments is disliked by referees, editors and readers. Succinct points score better than tedious rationalizations and ramifications that go down to minutiae. Small points will be understood by inference if the main points are clear and stated with due emphasis in the right context.

So before you draft the Discussion, list the main points you want to make; these can often be used as subheadings, making the structure of the text explicit.

The making of a good Discussion

The best plan is to identify the most salient points (five or fewer if possible) that emerge from your results. Deal with each clearly, discussing it in relation to other findings and in the context of your hypothesis. Put them under separate subheadings.

Marshalling your evidence and arguments in this way will allow you to draw the clearest conclusions and reiterate your message at the end of the Discussion. If your results have real substance, some of them should be able to stand alone as self-evident, needing little if any discussion. Very few papers have more than a single message to impart.

It is best to round off with a general statement. It is seldom helpful to say more than a few words about where your research might now lead, unless you are reporting a truly major breakthrough (which is very rare) or the findings call for a new research direction. Your paper will not be the final statement on the subject; everyone knows that 'further investigations will be needed'.

1.7   Acknowledging Others

We must always acknowledge the source of funding that made the work possible; to omit this is bad practice and will not help you to obtain further funding. You must also include in your Acknowledgements all the people, within reason, who have contributed materially and intellectually to the paper, or have gifted materials. Others will have criticized your drafts, and there may have been gifts of materials from different sources. It is often difficult to decide where to draw the line – but few if any authors need to go so far as to acknowledge the safety officer or security personnel in their laboratory!

Each author's contribution is intimated in no more than a sentence under Authors' Contributions. When many authors have made substantially the same contribution to the work, they should be mentioned together in the same sentence. Again, brevity is paramount.

1.8   The Relevant Literature

You should have gathered the References as you drafted the paper. Modern software packages such as Endnote have made this job far easier than it was in the past. But you should review the list and see that it includes only the most relevant publications. It should not aim to be comprehensive unless you are writing a review article. Remember that your readers are probably as knowledgeable as you on the subject and already know the relevant literature.

The style of the reference list will depend on the journal to which you will submit the manuscript, and most software can adapt the style without much extra work. Always check the required order of authors, title, journal, volume number and so on. Each reference has to be set out with the right spacing, correct font, exact punctuation, etc.

1.9   Missing Components

I have not forgotten the *Title*! As I said in the Introduction, the order in which a draft is produced is not the order in which it is finally presented. Imagine a newspaper editor has 'put to bed' the next day's copy. The thing he/she is likely to do last is to decide the headline – after all, a new sensational story could break at the eleventh hour. Then and only then can the decision be made as to what will appeal to the potential reader, hopefully a purchaser. Similarly, you will be able to produce the best title (headline) for your paper after you have finished rather than before you start drafting the manuscript. One type of headline that is frowned upon by editors is the 'declarative' title, which spills the beans – gives the answer to the question being researched, e.g. 'Very high testosterone levels in man increase the incidence of double-tailed sperms' – that might be enough for many readers, who will not bother to read the evidence presented in the paper. A good title, like a newspaper headline, should catch the attention and draw the reader into assessing the paper for what it contains.

Let us now return to the Abstract. You will find it much easier to write an Abstract when your first draft is otherwise complete. By then you have put everything in place in the article and can distil its real core. The advice is therefore to prepare the Abstract as the penultimate job in paper drafting, the final job being to decide the Title.

The next chapter will explore a typical paper. It will inform you as to how publishers like to set out articles in sections, in line with modern conventions. There is little diversity in how papers appear in different journals from different publishers. There seems to be quite strict conformity. This makes life easier for publishers, but often makes the scientific and medical literature look dull and tedious.

This is another reason for writing short, succinct papers that set out their messages clearly. For editors, it is a joy to read one of these when most of the other submissions are comparatively ponderous! For this reason Appendix 1.1 presents an actual case of how a verbose passage from a paper can be reduced to almost half its original length without loss of information, while also making it both easier to read and clearer to understand.

Appendix 1.1 Word Reduction

Here are two examples of an Introduction. The first (unpublished and prior to editing) version is overlong and lacks clarity.

Draft Version

Polyamines, such as putrescine (PUT), spermidine (SPD) and spermine (SPM), are polycationic compounds, and known to be widely distributed in every living organism. Previously, they have been reported to play an essential role in the cell proliferation. Recently, the biological actions of polyamines have been studied at the molecular levels, and these compounds have been suggested to be connected with the modulation of chromatin structures, the transcription and translation of the genes and the stabilization of the DNA as well as the functions of specific cellular proteins (Igarashi and Kashiwagi, 2000; Igarashi and Kashiwagi, 2010). Further studies have shown that polyamines can preferentially bind to the GC-rich regions of DNA and RNA, and the effects of these compounds on the GC-rich region of the DNA in a cell-free system have also been suggested to contribute to their *in vivo* effects (Igarashi *et al.*, 1982; Watanabe *et al.*, 1991; Yuki *et al.*, 1996). On the other hand, the previous studies have suggested that polyamines may be implicated in mental disorders (Fiori and Turecki, 2008), and also shown that polyamines can protect neurons against mechanical injuries, neurotoxic insults and ischemic damage (Clarkson et al., 2004; Ferchmin et al., 2000; Gilad and Gilad, 1999). Specifically, the genetic variants in the polyaminergic genes have been suggested to be associated with the psychiatric conditions, thereby proposing a possible connection between polyamine metabolism and mood disorders, such as anxiety, depression and attempted suicide (Fiori and Turecki, 2008). Further studies have provided evidence for suggesting the possibility that polyamines can probably contribute to the adult neurogenesis, the aged-related hippocampal neurogenesis and the learning and memory functions (Liu et al., 2008; Malaterre et al., 2004).

Previously, polyamines and their metabolizing enzymes have been reported to be localized in the different region of the brain or the different types of the cells, and therefore it seems possible to consider that polyamine synthesis and storage may occur at different locations in the brain [Bernstein, 1999]. On the other hand, polyamines have recently been shown to be preferably accumulated in astrocytes, thereby suggesting a possible role of polyamines in the regulation of the glial network under normal and pathological conditions (Benedikt et al., 2012). These findings are considered to propose the possibility that polyamines may be able to cause the modulation of neuronal cell function as a consequence of acting directly on glial cells in the brain, but little is known about the biological or the physiological actions of polyamines on the glial cell functions and metabolism. On the other hand, neuroactive 5α-reduced steroids have previously been reported to enhance the ability of C6 glioma cells to produce brain-derived neurotrophic factor (BDNF) through the promotion of their differentiation, thereby playing a putative role in protecting and reviving the functions of neuronal cells as well as maintaining the integrity of neural network in the brain (Morita et al., 2009; Morita and Her, 2008). Furthermore, the neurosteroid-mediated differentiation of the glioma cells has also been suggested to induce the enhancement of glutamate transporter GLT-1 gene expression, and therefore speculated to reduce the excitotoxic damage to neuronal cells as a consequence of facilitating the removal of glutamate from the brain tissue (Itoh et al., 2013). Based on these previous findings, it seemed possible to hypothesize that polyamines might cause the modulation of neuronal cell function probably through the enhancement of BDNF production in glial cells, which might be closely connected with the neurosteroid-mediated differentiation of glial cells in the brain. Then, as the first step for verifying this hypothesis, the direct effects of polyamines on 5α-R gene expression in rat C6 glioma cells were examined to obtain further evidence for suggesting their potential abilities to stimulate the biosynthesis of neuro-active 5α-reduced steroids, which can promote the differentiation of glial cells, thereby resulting in the enhancement of their potencies to enhance the BDNF production in the brain.

(644 words)

What follows shows how this Introduction can be made *clear and more succinct* using far fewer words.

Final Version

Polyamines, putrescene (PUT), spermidine (SPD) and spermine (SPM) – poly-cationics widely distributed in nature – are involved in cell proliferation, preferentially binding to GC-rich regions of DNA and RNA. They seem to modify chromatin, transcription and translation of genes, and are involved in the stabilization of DNA and the functioning of certain proteins (Igarashi and Kashiwagi, 2000; 2010). Their effects on nucleic acids in cell-free systems seem to correspond with those *in vivo* (Igarashi *et al.*, 1982; Watanabe *et al.*, 1991; Yuki *et al.*, 1996).

Polyamines have been implicated in mental disorders (Fiori and Turecki, 2008), possibly protecting neurons from mechanical, neurotoxic and ischemic damage (Clarkson *et al.*, 2004; Ferchmin *et al.*, 2000; Gilad and Gilad, 1999). Variants of polyaminergic genes may be associated with psychiatric conditions, indicating a connection between their metabolism and mood disorders, e.g. anxiety, depression and attempted suicide (Fiori and Turecki, 2008). Polyamines are involved in adult neurogenesis, age-related hippocampal neuro-genesis, learning and memory (Liu *et al.*, 2008; Malaterre *et al.*, 2004).

Polyamines and their metabolizing enzymes occur in different regions of the brain, where their synthesis and storage probably take place (Bernstein, 1999). Polyamines might preferentially accumulate in astrocytes, suggesting involve-ment in regulating glial networks under normal and pathological conditions (Benedikt *et al.*, 2012). They may modulate neuronal functioning by acting directly on glial cells, but little is known.

Neuroactive 5α-reduced steroids may enhance the ability of C6 glioma cells to produce brain-derived neurotrophic factor (BDNF) by promoting their differentiation, thereby protecting and reviving the functions of neuronal cells as well as maintaining neural network integrity (Morita *et al.*, 2009; Morita and Her, 2008). Neurosteroid-mediated differentiation of glioma cells may enhance glutamate transporter GLT-1 gene expression, reducing excitotoxic damage by facilitating glutamate removal from brain tissue (Itoh *et al.*, 2013).

Thus, we hypothesized that polyamines modulate neuronal functioning by enhancing BDNF production in glial cells, possibly connected with neurosteroid-mediated differentiation. In a pilot experiment, the effect of polyamines on 5α-R gene expression in rat C6 glioma cells examined their ability to stimulate neuroactive 5α-reduced steroid biosynthesis, which promotes glial cell differentiation thereby enhancing BDNF production.

(354 words; a 45 per cent reduction)

2   The Typical Scientific Paper
A Published Paper with Annotations

A research article is the *final product* of an investigation. It is a report that tells the world what you have done and found. On this account, it must be presented in the best possible way so that its message is to the point, clear and succinct.

2.1   Introduction: Types of Paper

There are several types of scientific paper:

- The **primary research paper**, which will be the focus of the present chapter.
- The **review article**. This provides an overview of the literature on a specialized topic, giving a balanced assessment of the current position. The emphasis is usually on recent advances and current controversies, but older publications (perhaps dating from the nineteenth century) may also be cited. Review articles are often written by experienced scientists with recognized expertise and may be commissioned by journal editors. A good review article is invaluable as a source of references, especially for newcomers to the field.
- The **preliminary communication**. This type of article contains recent data that are highly relevant to some special field. It contains insufficient information for a primary research paper, but it points to an important new development within that field and paves the way for a full article. It can be a way of 'getting a foot in the door' and it can also be a useful updating device in a fast-moving area of research and discovery. Some specialist journals deal exclusively with papers of this type (e.g. *BioScience Reports*).

- The **short communication**. Unlike a preliminary communication, a short communication must be complete within itself. Like a full research article, it imparts information that can stand up to scrutiny, but its topic will not necessarily develop much further in the foreseeable future. Alternatively, it might convey useful supplementary information obtained after a major paper was published, so it is the 'icing on the cake'. In recent years there has been the introduction of the **prepublication** – a short article that precedes a paper that will soon give the details in full. In effect this is 'to get your foot in the door' as soon as possible; it is a way of trying to ensure 'prior claim' to an emerging answer to a particular research question.

- The short general **'magazine'** article is seldom divided into the formal sections discussed in Chapter 1, and might not even have references. Its purpose is to draw attention to a hot topic in a field, often in non-technical language. It provides a quick update of developments so that interested readers can search the formal literature for more learned articles and discourses.

- **Technical reports** are papers that might or might not reach publication. For example, they can be prepared for discussion within a company, or they can lead to patent applications, instruction manuals, etc.

- **Abstracts** can be published in conference proceedings and/or relevant journals. They are more important to some scientists than others. Practising clinicians often write abstracts to alert their colleagues to findings that will not and perhaps cannot be followed up and published as full research papers. Scientists, however, normally follow up their meeting abstracts with full research papers in a learned journal. In some cases, abstracts written for conferences are considered as publications to be listed in one's CV, but they cannot substitute for full articles.

- **Other types of paper**. There are specialized types of paper that achieve a very specific purpose, such as a **'case report'** in a medical context. There are releases from companies giving information on new products or methods, and assessment papers that compare these products with one another. Journals can include biographies and obituaries, comment columns and so forth. The business of writing for the general public through the media by journalists is an entirely different matter, and it is here that a mixed bag of expertise is found. The result is often seen when learned papers indicating a significant advance in a field sometimes become misconstrued when reported through the mainstream press.

2.2   The Primary Research Paper

We will now dwell mostly on the primary research paper. For convenience, this will be illustrated using a research paper of average length from my own group.

You will recall from Chapter 1 that each journal has its own style of presentation. Therefore, when you decide to which journal you will submit a manuscript, you need to know its format and style. Details of preparation and submission procedures are given in the *Instructions (or Guidelines) to Authors*, which are found within the online submission systems for each journal and publisher. If you do not follow these instructions exactly, most journals will return your manuscript or send a message asking you to upload a new version that conforms to the required style. During the past 30–40 years, editors and publishers have become increasingly insistent that authors take full responsibility for the presentation of their articles. Since the advent of online publishing they have insisted even more on precise conformity to the journal's instructions for every submitted manuscript. A paper that arrives in an editor's email sometimes has a totally different format from the one required, in which case the immediate suspicion is that the article had been sent to another journal and been rejected. Without amendment, it has been sent to this new one. If it had been sent with a covering letter explaining its history to the editor, it might have been assessed as to its suitability, and if approved, the authors might easily have been asked to resubmit the paper in the requisite form.

Notwithstanding that the onus of presentation is entirely on the author, the appearance of the published paper in the journal differs in several respects from that of the accepted final manuscript. Editors and print managers upload the typescript into a template that deals with such details as the positioning of figures, division of the text into columns (authors are seldom asked to set out their manuscripts in columns), preparation of footnotes (e.g. for the corresponding author's contact details), the fonts and sizes for titles and subheadings, and so on. It is therefore unnecessary, and inappropriate, to prepare and submit a manuscript that looks exactly like the finished product. This is why figures, tables and other such items are normally appended at the end of a paper, or submitted as separate files.

2.3   An Annotated Research Paper

The following annotated paper illustrates and explains many of the above-mentioned points. I make no apologies for having used one of my own papers simply because it has been conveniently at hand on many previous occasions when I have been delivering one of my courses to students and staff at institutions all around the world.

The paper is presented page by page in its final format. Explanatory notes appear as near as possible to the section being discussed. These notes are not comprehensive as each section of a paper will be discussed in full detail in subsequent chapters, where instruction is given about how to draft and improve your paper as it takes shape. The present chapter will be most useful for readers with little or no experience of writing papers. For most scientists, a glance at a paper from the journal chosen for submission is by far the quickest way to get the mind-set needed to proceed with a draft. Look at other papers in the literature to become familiar with their general presentation; this is like seeing a picture, explaining things better than the proverbial thousand words.

Available online at www.sciencedirect.com

ELSEVIER

Cancer Letters 227 (2005) 141–152

www.elsevier.com/locate/canlet

Integrity and stability of the citrulline–arginine pathway in normal and tumour cell lines

Denys N. Wheatley[*,1], Ruth Kilfeather, Alison Stitt, Elaine Campbell[2]

BioMedES, Hilton Campus, Hilton Place, Aberdeen AB24 4FA, UK

Received 7 October 2004; received in revised form 2 December 2004; accepted 7 January 2005

Abstract

Arginine catabolizing enzymes have been used on cancers for over 60 years. In the last 5 years the ability of arginine catabolizing enzymes, not only to inhibit proliferation, but to kill tumour cells has been reinvestigated. Selectivity of action lies in the inability of many tumours to circumvent arginine deprivation by recycling precursors through the urea cycle. While this offers an immediate window of opportunity to treat, e.g. melanomas and hepatocellular carcinomas (HCC) that have poor citrulline converting ability, it is possible that the deprivation can be applied to many other types of cancer. The problem of deficiency of the urea cycle enzymes in a wider range of normal and malignant cell lines has been addressed, and shown to be variable throughout several different tumour types. We also need to know how fickle recycling enzyme activity can be in both normal and tumour cells, and found to be remarkable stable. This latter point bears in the amino acid (arginine) deprivation protocol because it has already moved into the clinic. Initial findings on a named-patient basis have been encouraging, and the development of a new rational approach to the systemic treatment of melanomas, HCCs and leukemias seems imminent. This is the more attractive because arginine deprivation protocols can also 'stage' tumour cells for combination therapy in cases where they might not be killed outright by deprivation alone.
© 2005 Elsevier Ireland Ltd. All rights reserved.

Keywords: Arginine metabolism; Catabolism; Enzymes; Citrulline; Arginino-succinate; Urea cycle; Melanoma; Hepatocellular carcinoma; Cancer therapy

1. Introduction

* Corresponding author. Tel.: +44 122 427 4173; fax: +44 122 427 4179.
E-mail addresses: wheatley@abdn.ac.uk (D.N. Wheatley), elaine.campbell@st-andrews.ac.uk (E. Campbell).
[1] Funded by: grants from the Synos Fondation (Switzerland), Grampian University Hospital Trust Endowments Fund, and the 'Friends of Anchor' (Balmoral) Cancer Group, Scotland.
[2] Present address: Department of Biological Sciences, University of St Andrews, Bute Street, St Andrews, Fife, Scotland.

0304-3835/$ - see front matter © 2005 Elsevier Ireland Ltd. All rights reserved.
doi:10.1016/j.canlet.2005.01.004

In view of a number of related studies from other laboratories on the arginine degrading enzyme, arginine deiminase [1–5] and the advance of this enzyme into phases I and II clinical trials [6], notably in pegylated forms [7,8], the potential of arginine deprivation as a new modality in cancer therapy has to be seriously entertained.

Notes on the Format and Content of the Typical Primary Research Paper

Headings at the top: The publisher adds the heading, giving the reference of the paper in the journal, the company and its logo. The author leaves this to the company.

Title area: Fonts and sizes will be dealt with by the publishers, but the affiliations after the authors' names are as superscripts (see Footnote). Below this the publishers adds details of the dates of submission, revision and acceptance. This is useful if there is a priority dispute (i.e. someone else claims to have published similar findings and conclusions at an earlier time).

Abstracts: These usually have to be no more than a certain number of words (e.g. 150 or 200). Put the main results here, without detail, and with no more than a sentence of introduction or discussion. Some journals ask for structured abstracts (dealt with in Chapter 7). Citations are only included if absolutely necessary. Make sure the 'take home message' is prominent. Keywords (usually up to 6) – sometimes written unnecessarily as Key Words – are given below the abstract. Unfamiliar abbreviations are often placed below Keywords, usually when there are many, rather than having to do this in the text.

Note that there is a line under Keywords. A single column set-up has been used so far by the publisher, but thereafter the paper is set out in two columns. This arrangement is now common, although many journals do not use this column layout. (The two-column layout is used because it is easier to scan 5–6 words per line than 10–12 words per line, arguably for faster reading.) However, it has its drawbacks (see Chapter 9) regarding how figures, tables and other illustrations fit in a column.

Footnote: At the bottom of the first column is a footnote created by the publisher that gives the contact details of the corresponding author, the funding sources and other relevant information. In some journals, the footnote information may be placed elsewhere, e.g. under the authors' names. The words under the footnote give details of the copyright holder (in this case Elsevier), followed by a 'DOI' (digital optical identifier), a unique number that allows the paper to be found quickly on the journal website and by literature search engines.

Introduction: You will see that each paragraph is indented by about three spaces, whereas my notes have no indentation at the start of each paragraph; this is the publisher's preference. References in the text are the numbered 'Vancouver' style, e.g. [14], and will be listed in numerical order in the Reference section. Alternative referencing systems will be discussed Chapter 8. The Introduction continues on the second page, in the left-hand column.

142   *D.N. Wheatley et al. / Cancer Letters 227 (2005) 141–152*

In addition, work with other enzymes (arginase and arginine decarboxylase) emanating from other laboratories [9–13] also shows promise. Differences between the effects of these arginine-catabolizing enzymes exist, although stricter comparisons need to be made and optimal conditions found [14]. Some evidence shows that deprivation induced by arginine deiminase has toxic sequelae because ammonia is released [15], which affects both normal and tumour cells. Furthermore, the other product of the deiminase reaction is citrulline in equimolar amounts to the arginine substrate, and therefore cells with intact urea cycle should be able to escape arginine deprivation by recycling this citrulline.

Our recent data clearly shows that, where tumour cells cannot handle citrulline because of the failure of arginine and citrulline interconversion via argininosuccinate synthetase (ASS) and argininosuccinate lyase (ASL), they are seriously disadvantaged by arginine deprivation [16], and therefore should be excellent targets for selective destruction. Although the conversion of citrulline to arginine occurs, except where kidney function has been compromised, demonstration of the inability of tumour cells per se to use citrulline means that a constant plasma level of arginase or arginine deiminase can prevent citrulline recycled to arginine in the kidney becoming available to the cells again. Hence, most tumours will preferentially regress and die where they fail to move out of cycle [11]. Cells stuck in cycle remain vulnerable to low doses of cycle-dependent cytotoxic agents [14].

Sensitivity of tumours can quickly be determined by appropriate testing for argininosuccinate synthetase (ASS) activity, or by simply testing their ability to grow in vitro in citrulline. Because most melanomas and hepatocellular carcinomas (HCC) are deficient in ASS [2,17,18], treatment with enzymes catabolizing arginine looks particularly promising, especially for two tumours types that are particularly malignant. However, our investigations on a selection of different tumour types [16] suggest that it may be too early to generalise from and about melanomas and HCC data, and there seems to be no logical reason why tumours of these two particular types should seem almost universally deficient in the same urea cycle enzymes [2,17,18]. Our present report contains data on many tumour and normal cell types, and reveals some unexpected findings with regard to citrulline metabolism.

2. Materials and methods

2.1. Cell culture

Cells from colleagues and a wide range of suppliers were grown in RPMI 1640 medium (Life Technologies, Paisley, UK) at 37 °C in a humidified atmosphere of 5% CO_2. The medium was formatted such that it could either contain arginine, usually at 1 mM or be arginine-free. Stock cells were grown in the presence of 10% fetal calf serum, but treated cultures had a reduced level (see below). Cell counting involved a Coulter electronic particle counter (Model ZM; Coulter Electronics, Luton) or a Nucleocounter (Chemometec, Denmark). Counts are representative of the normal cell population, within standard parameters, and unless otherwise stated cannot be taken as an indication of viability. All cell cultures were routinely screened on arrival and immediately prior to experimental use for four types of mycoplasma, using the Mycoplasma Screening Kit of Roche (Lewes, UK), according to the manufacturer's instructions.

2.2. Experimental design

Cells were seeded at 50,000 cells per well in 12-well plates depending on generation time or unless otherwise stated, and allowed to settle in RPMI medium containing 1 mM arginine and 5% dialysed fetal calf serum (FCS) overnight. Prior to treatment, cells were washed with pre-warmed arginine-free medium (hereafter AFM) and fresh AFM introduced to which, arginine, argininosuccinate, or citrulline was added at 1 mM, as and where appropriate, or at other concentrations as noted in the appropriate places. In the case of L1210 cells, a suspension culture line, cells were treated immediately on being resuspended at the desired concentration in arginine-free medium to which arginine, argininosuccinate or citrulline was added as desired. These compounds were obtained from Sigma (Poole, UK) in the purest available form. In the case of argininosuccinate, the purity level was > 85%.

Materials and Methods (M & M): The next major heading of the paper is usually subdivided and numbered. Thus, the subsections are 2.1, 2.2, 2.3, ... Further subdivisions of 2.1 would become 2.1.1, 2.1.2, 2.1.3, ... Occasionally this can go another step further, 2.1.1. leading to 2.1.1.1, 2.1.1.2, ... This last degree of subdivision seems to go too far. A better option is to go back to the shorter numbering (say up to 2.1.1) and then for further subdivision the options include indenting the heading and perhaps changing the font, as follows:

2.1.1 This could lead to:

Chemicals

Cell culturing

Animals

These subsections could also be listed as (a), (b) and (c). The advice is simply to try to avoid having too many subsections.

Traditionally, the first part of M & M deals with the chemicals, cells, animals, etc., and subsequent subsections are more devoted to the technical details of the experimentation. Note the sources of substances, which is required because a chemical can differ from supplier to supplier, usually in purity; this can even happen with different 'lots' or 'batches' from one supplier.

Acronyms and abbreviations: At the first appearance of some words, such as an organization, a unit or a chemical, often occurring in M & M and thereafter throughout the text, they should be given in full followed in brackets by the acronym or abbreviation, e.g. 'arginine-free medium (hereafter AFM)', although I am stressing it by adding the word hereafter, which is unnecessary. Acronyms that are in common use need not be put in full e.g. DNA, WHO, GMT.

The SI system of units is used in almost all papers. There are accepted abbreviations for these units that need not be spelt out. However, you must be consistent, e.g. 'minutes' is always 'min', not 'mins'. Concentrations of pure chemicals in solution are better given in molar terms (e.g. 1 mM) rather than as grams per litre (g/L). There are many inconsistencies in the way different journals ask for units to be stated. Here I will mention just three, as this issue will be discussed in later chapters: for example, 1 mM can also be 1mM (no space), but 5 % should be 5%, and it is 37°C, not 37 °C.

In the right-hand column, the M & M section ends with the statistics subsection, which is necessary in the vast majority of papers in science. Traditionally it seems to have always been tagged on to the end of section 2 (as in this case).

D.N. Wheatley et al. / Cancer Letters 227 (2005) 141–152   143

Cell counts were made usually at daily intervals over 3 days for fast growing cells and over 5 days for slower growing cells. Cells were characterised according to the following system:

Type A. Grow only when arginine is present;
Type B. Grow on arginine or argininosuccinate, but not citrulline;
Type C/As+. Grow on citrulline and argininosuccinate;
Type C/As−. Grow on citrulline, but cannot grow on argininosuccinate.

Cycloheximide (Sigma, Poole, UK) was prepared in Dulbecco's phosphate buffered saline as a stock at 1 mg ml^{-1} and diluted 100-fold for treatment of cultures to inhibit protein synthesis. Arginase was prepared as before by Dr Ikeomoto (Kyoto, Japan), with a specific activity close to 1100 U mg^{-1}; it was added where required at 1 Unit ml^{-1} medium.

2.3. Radioactive incorporation data

Cells were grown in the presence of 0.4 mM arginine and citrulline containing 1 μCi ^{3}H-arginine or ^{3}H-citrulline, respectively. On sampling, the cells were washed 3× in ice-cold PBS, after which 0.6 N PCA was added to precipitated the proteins. One hour later the supernatant was removed and duplicate 100 μl samples from this acid-soluble fraction (the cytoplasmic pool) were mixed with 2 ml of PicoFluor™ 40 (Packard, Downers Grove, Ill, USA) and analysed for ^{3}H activity in a 1219 Rack Beta liquid scintillation counter (Wallac, Gaithersburg, MD, USA). After removal of

the supernatant, the PCA precipitated material was dissolved in 1 N NaOH for 1 h at room temperature, and then thoroughly mixed. Duplicate 100 μl samples were taken and measured in the scintillation counter as described above, being the incorporation proportional to the pool that had entered the proteins. ^{14}C-argininosuccinate was kindly prepared for us by Dr Oliver Musgrave, using ^{14}C-arginine as the substrate (Amersham, UK), with little loss of label.

2.4. Statistics

Each experiment reported here had been repeated <3 times. Data were collected each time-point from a minimum of three samples, and the mean and one SD calculated and plotted. Probabilities have been worked out only where necessary, the majority of situations being very clear as to whether or not cells would grow with particular substrates.

3. Results

We have studied at least 10 normal and over 50 malignant human cell lines, not all of which are included in Tables 1 and 2. These tables include cell types not less than three times, and data are available on another 11–12 lines that have so far been tested only a couple of times.

3.1. Normal human cells

Some of the 'normal' cell types came from our considerable fibroblast stocks grown up to nine

Table 1
Normal human cell cultures: categorization in terms of their ability to utilize citrulline, argininosuccinate or arginine

Cell type	Species	Details	Type A	Type B	Type C	
					AS+	AS−
HUVEC	Human	Endothelial	+			
HMVES	Human	Endothelial	+			
FF6/1	Human	Foreskin fibroblast			+	
FF6/9	Human	Foreskin fibroblast			+	
LSF/9	Human	Adult skin fibroblast			+	
HKP3	Human	Prox. renal tubule epithelium	+			
HEK	Human	Prox. renal tubule epithelium	+			
HK/(VR)[a]	Human	Prox. renal tubule epitlelium (primary)	+			

[a] First passage of a normal human kidney proximal renal epithielial cells from a kidney resection.

The Results section: Here, in section 3, tables and figures are used to present much of the information germane to the hypothesis being tested. This saves having to spell out details in the text, saving space as well as making things far easier and clearer for the reader to comprehend. There is a tendency for authors to put too much information surrounding the data into this section. It should be as factual as possible, and not a place for any general comments. The bolder and more unadorned the statements and the information are, the better the reader can grasp the main points of evidence being worked on in testing a hypothesis. He or she hopes that a lucid discussion on the relevance of the findings will be found in the next section; giving a full ongoing explanation while recording the experimental data in the Results section is to be avoided.

Data being presented in the Results section should be in as logical an order as possible. It may consist of many different pieces of (experimental) evidence, but like a court case, the presentation has to make for 'a convincing story'. The difficulty here is in knowing exactly what to put in as overload can be counterproductive. A century ago papers were long and may have included data collected over years, making them like lengthy laboratory notebooks. Nowadays papers are shorter and much more focused on answering a specific question, testing a simple hypothesis.

Note that when a paper has been accepted for publication, the final proof will come back to the author for checking. It is surprising how often mistakes are missed. Look at the fourth line of the first paragraph of the Results. The word 'tested' has been omitted after 'types'. Another example is in the second line of 2.4, where the authors intended '3 times', but '< 3 times' was written by mistake. It is worth getting an 'independently minded' person to seek out errors like this before a paper is submitted.

144   D.N. Wheatley et al. / Cancer Letters 227 (2005) 141–152

Table 2
Human malignant/transformed cell lines: categorization in terms of their ability to utilize citrulline, argininosuccinate or arginine

Cell type	Species	Details	Type A	Type B	Type C	
					As+	As−
A375	Human	Melanoma		+		
G361	Human	Melanoma				(+)
MEWO	Human	Melanoma		(+)		
MeME/3	Human	Melanoma	+			
HL428	Human	Lymphocytic leukaemia				+
JURKAT	Human	Lymphocytic leukaemia				+
MOLT4	Human	Lymphocytic leukaemia				+
Wil2	Human	Lymphocytic leukaemia	+			
CEM	Human	Lymphoblastic leukaemia	+			
CEM-DR	Human	Lymphoblastic leukaemia	+			
Namalwa	Human	Lymphoblastic leukaemia				+
TK6	Human	Lymphoblatic leukmia	+			
Wi-L2-NS	Human	Lymphoblastic leukaemia	+			
Skut1B	Human	Uterine leiomyosarcoma	+			
Khos-NP	Human	Osteogenic sarcoma				+
SaOs-2	Human	Osteogenic sarcoma		(+)		+
U2OS	Human	Osteogenic sarcoma				+
Ca Ski	Human	Cervical carcinoma				+
HeLaᵃ	Human	Cervical carcinoma				+
HeLa (USA)ᵃ	Human	Cervical carcinoma				+
HeLa (S3)ᵃ	Human	Cervical carcinoma				+
MCF7	Human	Breast adenoma				+
A549	Human	lung carcinoma	+			
A431	Human	Myeloma				+
Hep3b	Human	Hepatocellular carcinoma				+
HepG2	Human	Hepatocellular carcinoma				+
WiDr	Human	Colorectal adenoca'oma				+
MOLT4	Human	Lymphoblastic leukaemia				+
ACHN	Human	Renal carcinoma				+

( ) denotes weak use of the appropriate substrate for categorization under this type.
ᵃ From separate stocks obtained over twenty years from three different sources (see Table 3).

passages from primary explants from patients undergoing plastic surgery or circumcision. We have never had a response different from that shown in Table 1; that is, every fibroblast culture examined from first subculture to passage nine has invariably been Type C (As−), which means that the cells can grow on citrulline, but never use argininosuccinate (AS). We are talking here of > 10 separately isolated fibroblast cultures. The first and ninth passage of foreskin fibroblasts from one stock (FF6) have been included in Table 1, designated FF6/1 and FF6/, and were confirmed as Type C by fibroblasts from the ninth passage of LSF adult skin (LSF/9), as well as those reported in a previous communication [16]. We can confidently state that normal human fibroblasts (adult and neonatal) show stable expression of the ASS/ASL

pathway that reflects both high active and strong coupling, because fibroblasts grown in citrulline at 1 mM at close to the rates in arginine.

Our laboratory has been receiving a frequent supply of fresh endothelial cells, used in our angiogenesis assay at passage 2 or 3 [19] (Bishop et al., 1997; cells provided by BioWhittaker, Clonetics or TCS CellWorks, Botolph Clayton, Bucks, UK). Both human unbilical vein (HUVEC) and human microvascular endothelial (HMVEC) cultures consistently showed absolute dependency on arginine for growth (Type A), with citrulline and AS affording no sign of rescue.

Perhaps more surprising was our finding that two sets of cultures of human proximal renal tubule epithelial cells (HEK and HKP3) were also as totally

Tables: The page shown here is largely taken up with a Table. Table 2 deals with a large number of cell lines from different animals, their nature (whether malignant or not), and in which of three different types they belong. This amounts to a considerable body of information which can be easily perused. Some explanation of the data is given in the text; however, the data are not fully reiterated in the text, only the most salient points. All the relevant information is presented in the most succinct way, keeping the paper as short as possible, and making the results easily assimilated by the reader.

No emphasis has been given to the title by the publisher of this Table, but sometimes its title can be in bold font, following immediately after the word Table (e.g. **Table 2: Human malignant...**). In this case the publisher has separated the word Table and put the title on the next line. The words do not stand out as they are in small font and not in bold (it is not advisable to put the title beneath the table). This should have been changed at the proofreading stage, as with similar cases throughout the paper (e.g. figure legends). The choice should be up to the author, but publishers use templates that make these differences in presentation. There are small inconsistencies here; Table is written out in full above the table, but figure legends are beneath the illustration, using the abbreviation Fig. It is best to leave a space after the figure number. Some authors prefer, e.g., Figure 1 – legend, others prefer Figure 1: legend, and so on; whatever the choice, the chosen form must be consistent.

Titles can sometimes be full sentences in themselves, in which case a full stop/ period can be used at its end, otherwise leave one out. Only brief slim tables would fit into the narrowness of a column on a two-column page. Here the table is spread across both columns. Footnotes to a table should be tight against the bottom of the table's frame and preferably of a slightly smaller font. Some tables are published with the gridlines showing, but the content can often appear clearer if they are omitted. They are more often used when tables are lengthy.

Positioning tables and figures in the relation to the text is also an issue to consider. Authors have to put up with the print manager's preference, but tables and figures need to be as close to their relevant text as possible. If the print manager's arrangement seems unsuitable, this could be sorted at the proof-reading stage. The reason for this complication is because most journals ask for figures, tables and other illustrations to be sent as separate files at submission, but the author can indicate where they ought to go in the text.

D.N. Wheatley et al. / Cancer Letters 227 (2005) 141–152   145

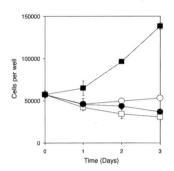

Fig. 1. G361 growth in arginine-free medium (AFM) to which arginine and its precursors was added. G361 cell growth over 3 days in AFM to which was added 0.4 mM arginine—positive control (■), nothing—negative control (□), 0.4 mM citrulline (○) or 0.4 mM argininosuccinate (AS) (●).

dependent as endothelial cells on arginine for growth (Type A). This was in agreement with primary cultures of human proximal tubular epithelial cells provided by Dr Vicente Rodero (see Section 4 for a careful consideration of these findings).

3.2. Human malignant and transformed cell lines

In Table 2, the cell lines have been loosely clustered into groups of their reputed tissue of origin. Of the four melanoma cell lines, G361 could barely use citrulline as a source of rescue from arginine deprivation (Fig. 1), and high concentrations were also employed to see whether some growth could be enhanced, but with no obvious improvement in growth success (data not shown). We had previously considered G361 to be Type A. On balance over many trials ($n \sim 7$), we now believe that G361 may retain very weak Type C characteristics, whereas A375 and MEWO showed consistent—but slow—positive growth in AS, quite unlike G361. This clearly distinguishes Type B from C. A second batch of G361 cells obtained from a completely different source gave identical results to the first batch.

When AS was provided at 1 mM in arginine-free medium along with 1 Unit ml^{-1} arginase, there was

no difference in kinetics from the G361 control cultures lacking arginase. This was also the case with A375 and MEWO cultures. However, the question as to whether the argininosuccinate preparation contained citrulline or arginine is still relevant, and will be considered more fully in Section 4.

The results with MeME/3 were also difficult to evaluate, but it seemed totally dependent on arginine, and is considered to be Type A. A further 14 melanoma cell lines are under investigation in a parallel study with primary cultures of melanoma cells from patients' tumours following resection (work in progress).

Of the five human lymphocytic cell lines tested, four were 'wild-type' (type C/As−) and only one differed, viz. Wil2 (Type A). Interestingly, the situation was reversed with lymphoblastic leukemias, in which five out of six were Type A, and only one (Namalwa) was Type C (As−). There has been less data on leukemias, and therefore the significance of this difference remains elusive at the present time.

Table 2 also includes many other tumour lines, most of which are wild-type, except for an osteogenic sarcoma (SaOs2; Type B) and a lung carcinoma (A549; Type A). Clearly a lower overall percentage of tumours are going to be deficient in ASS/ASL than originally suspected [16], and markedly lower than others have found in HCCs and melanomas [2], but clearly more data is required before any generalisation can be made.

With regard to the HeLa cell line, we held three different stocks which had remarkably different histories; all were Type C.

3.3. Non-human cell lines

It was important to follow cell types in other species in which more experimental work can be done in vivo. Table 3 shows again differences, but the exercise here was to type cultures sent to us from the four corners of the world specifically for this purpose. The data is compiled for four well-known types that have been in culture over many years, some of which were mentioned briefly in passing in Wheatley and Campbell [16]. The mouse melanoma line B16F10 has proved remarkably consistent, with all the four samples brought in from different laboratories designated (a)–(d) retaining an excellent ability to utilise

Figures also encapsulate a lot of information, and once again full details need not be reiterated in the text. Brief explanations of the data may be written, but comments on the significance of the findings should be deferred until the next section (Discussion). Figures are best kept as simple as possible; too much information makes it difficult for the reader to sort out what is what. The symbols conventionally used are open (○) or filled (•), using circles, squares, triangles, etc. Solid and dashed lines can then be used if greater distinction is needed. The axes of a graph should not be too disparate in length (a 4:3 ratio is best). Confidence limits or standard deviations are usually shown as error bars when the data points are averages of ≥ 3 values. Figure 1 in this paper shows a few error bars at 1 h, but at most of the other time points the error bars cannot be seen because they lie within the size of the symbol. The publisher places the legend beneath the figure. Whether experimental details need to be included in the legend can be controversial; this and other related issues regarding figures will be fully dealt with in Chapter 9.

The rest of the text of the Results section continues on this page. The subsections deal with different sets of data in a way that helps the reader to see the overall plan of the experimental series. This enables the facts to be presented as plainly and logically as possible. At this stage they are not discussed any further or in relation to other people's findings. The main object of this paper has been to provide comparative information on differences between normal and cancer cells regarding their metabolism of certain amino acids. These differences will then be the main subject of the next section, the Discussion.

Table 3
Non-human cell lines. Categorization of their ability to utilize citrulline, argininosuccinate or arginine, but with emphasis on the stability of expression in cultures from different sources on their characteristics in four of the six cell lines

Non-human	Species	Details	Type A	Type B	Type C	
					AS+	AS−
B16F10 (a)	Murine	Melanoma				+
B16 F10 (b)	Murine	Melanoma				+
B16F10 (c)	Murine	Melanoma				+
B16 F10 (d)	Murine	Melanoma				+
B16 F1	Murine	Melanoma				+
B16 F0	Murine	Melanoma				+
B16 B02	Murine	Melanoma				+
L1012 (a)	Murine	Lymphocytic leukemia				+
L1012 (b)	Murine	Lymphocytic leukemia				+
L1012 (c)	Murine	Lymphocytic leukemia				+
CHO (a)	Hamster	Ovarian carcinoma	+			
CHO (b)	Hamster	Ovarian carcinoma	+			
CHO (c)	Hamster	Ovarian carcinoma	+			
CHO (d)	Hamster	Ovarian carcinoma	+			
MDCK (a)	Canine	Kidney epithelial	+			
MDCK (b)	Canine	Kidney epithelial	+			
MDCK (c)	Canine	Kidney epithelial	+			
MDCK (d)	Canine	Kidney epithelial	+			
H35	Rat	Hepatoma			+	
McCoy[a]	Murine/human (?)	Fibroblast (?)				+

[a] The McCoy cells looked decidedly more like epithelial cells than fibroblasts. The possibility exists that this cell line was for a long time ago and is in fact a variant of the HeLa cell line; its categorisation agrees with that of Hela (Type C).

citrulline, almost as efficiently at 1 mM as arginine, the kinetics being indistinguishable on several occasions (generation time in both ~22–24 h; see Fig. 2).

Of particular interest were three further variants of the B16 line, namely B16 F0, B16 F1 and B16 B02, which exhibit different properties with regard to in vivo growth and metastatic potential [19,20]. Unexpectedly, all three retained ASS/ASL activity, characteristic of Type C (As−).

Turning to the mouse lymphocytic leukaemia L1210 cell line that we have frequently used [12], our stock has also been consistently Type C (As−). Three other L1210 lines from distant colleagues were also Type C (As−).

McCoy cells are included in Table 3, although as noted in the legend, these may be a contaminated HeLa line, and were also Type C (As−).

MDCK cells have a strong tendency to adhere to plastic culture-ware, and this was the case with our own stock as well as cultures from three other laboratories. Their appearance in cultures was very similar, being very flat characteristically pavement epithelial cells with large variations in individual cell sizes (areas). They were consistently Type A, failing to recover in citrulline and AS. The Rueber H35 rat hepatoma cell line, the only rat cell line, was Type C.

Finally, Table 3 includes a very old CHO stock line '(a)' compared with three others from disparate sources around the world (b–d). CHO has been our most assured Type A for several years, and it was reassuring that cells grown elsewhere over many years retained this same total inability to utilise citrulline and AS, although such a finding is less unexpected than with the lines above for reasons which will be explained in Section 4.

3.4. Attempted adaptation to citrulline in G361 cells; single step and 5-week gradual shift

Following either a 5-week gradient adaptation period, G361 cells were grown in RPMI arginine-free medium to which was added either arginine or citrulline at 0.4 mM from the start or in a gradient switch over from arginine to citrulline at a 20% increment per week.

The Results section continues: Little needs to be said here, the Results are presented in the same way as on the previous journal page. However, you will begin to appreciate the amount of work that went into producing this one primary research paper. It was the culmination of about three years of research. This is because over 70 different cell lines were included in the investigation, and these had to be sourced from colleagues all over the world to make the closest comparisons. For example, MDCK cells came from four unrelated laboratories, and all four proved to be type A. If they had been different in their typing, one would suspect that they were not all the same authentic kidney epithelial cell line or some had changed (e.g. adaptation to different media, or mutation).

The reference to Table 3 comes half way down the right-hand column. On reflection, although of no particular consequence, it might have been wiser to position Table 3 at the bottom of the page rather than at the top, which might have been corrected at the proofreading stage.

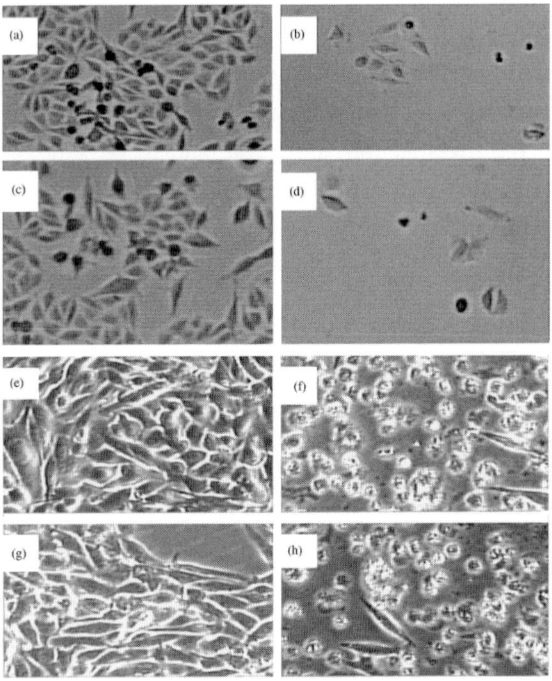

D.N. Wheatley et al. / Cancer Letters 227 (2005) 141–152   147

Fig. 2. Images of cells in media with and without arginine and its precursors. (a–d) A375 cells grown in the presence of 0.4 mM (a) arginine, (b) AFM only (negative control) (c) citrulline or (d) argininosuccinate. Phase-contrast, ×180. (e–h) B16 F10 cells grown in the presence of 0.4 mM (a) arginine, (b) AFM only (negative control) (c) citrulline or (d) argininosuccinate. Phase-contrast, ×180.

Fig. 3 shows growth in citrulline was virtually identical in both the adapted and parental strains, whether an immediate step or a 5-step gradient was used. Interestingly, growth in arginine was slightly depressed in the 'adapted' cells, for which we have as yet no explanation. Similar data to be reported later showing an absence of induction and no better utilisation of citrulline following long-term exposure was obtained with G361 cells from a separate source, as also with KHos-NP cells

(a human osteogenic sarcoma cell line) and L1210 cells, which were run in parallel studies (to be reported).

3.5. Uptake of labelled citrulline and arginine argininosuccinate

In growth experiments with precursors the assumption can too easily be made that they have ready access to the cell internum, i.e. no transport problems.

Figure 2 here is a composite; a collection of micrographs showing control cell cultures on the left and treated cultures on the right as pairs for comparison. The contrast in each image has been compromised by degradation during copying into this chapter, as it will already have been degraded when the original paper was being published. Each step in the procedure affects the clarity. Figures showing micrographs are normally lighter, but the contrast has to be adjusted to ensure the right balance. In preparing your own paper, you will need to pay attention to the matter – try to make the images as pristine as possible, but keep in mind the size of the files involved (in kB, MB and hopefully not GB!). Modern software can greatly improve images, but must not be used to tamper in ways that could lead to misrepresentation of the material.

Although definition, contrast and clarity are the main factors to consider when presenting figures of this kind, reproduction of Figure 2 in this paper has not been that successful compared with the original images. We have to acknowledge that the figure reproduced here for illustrative purposes falls short of an acceptable standard.

148   *D.N. Wheatley et al. / Cancer Letters 227 (2005) 141–152*

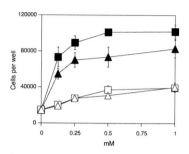

Fig. 3. Effect of 5 weeks exposure to citrulline in increasing levels from 0 to 100% on the growth on the G361 cells subsequently in arginine and citrulline containing medium (closed vs. open symbols, respectively). Unexpectedly 'adapted' cells grew less well when returned to arginine medium. Other symbols: 'non-adapted' cells grown on arginine (■), 'non-adapted' on citrulline (△); 'adapted' cells grown on arginine (▲), 'adapted' cells grown on citrulline (□).

turned off with cycloheximide ($10 \, \mu g \, ml^{-1}$), which suppresses protein synthesis by 90–95%; this is because a catabolic pool boosts the total acid-soluble pool in the case of arginine. It also shows that the very slow increase in the acid insoluble pool (macromolecules) above background was unrelated to proteins synthesis because it occurred in both cycloheximide-treated and control cultures (Fig. 5b). It also occurred the same way when arginase was

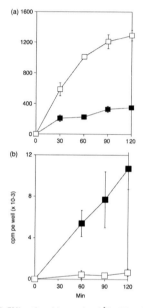

Fig. 4. G361 uptake and incorporation of ^{3}H-arginine and ^{3}H-citrulline into acid-soluble and acid-insoluble compartments from. (a) Uptake into the acid-soluble pool of G361 cells of ^{3}H-arginine (■) and ^{3}H-citrulline (△), each point being the mean of six samples, with the error bars showing ± one SD. Note the larger pool created by citrulline. (b) Corresponding incorporation into acid-insoluble material in G361 cells. The symbols are as for (a). Note the much smaller incorporation of citrulline (as arginine following conversion by ASS) into protein.

Although this can be true, few previous studies have checked this out. Here, virtually all experiments have been done with exogenous citrulline in comparison with arginine at concentrations (0.4 or 1 mM) that should easily support the G361 cell line from low seeding densities so that the supply of amino acid is in surplus and no limiting conditions would be encountered in a 3-day growth experiment. Citrulline had no problem accessing the internum of G361 cells (Fig. 4a), but the incorporation was only ~10% that of arginine incorporation into protein over the 2 h period of incubation (Fig. 4b).

Labelled-AS was transported into the acid-soluble pool of CHO (Fig. 5a), which were selected because non-transport of the substrate in this case would readily explain why these cells respond so negatively to the intermediate. The other cell line chosen was A375 because it can utilise AS, if only very slowly. Our data clearly show that this second cell line also had no problem with the uptake of AS into the intracellular pool (Fig. 5a), and therefore slow incorporation into protein reflects the slow conversion to an adequate concentration of radiolabelled precursors being present. Comparison with arginine in pools was more difficult unless protein synthesis was

The Results section continues: This paper has a wealth of data to report, and this page looks similar to previous ones. Note, however, that the figures retain the same style and are small in size to fit the columns. If the figures are presented in different ways, they are more difficult to follow than when a consistent pattern is maintained. Indeed, if similar control and experimental groups are reported in different graphs, it makes sense to use the same symbols for them in each of these graphs.

Although Figure 4 in the right-hand column is quite simple, some of the experimental details are given in the legend in this case because it would have been too burdensome to include them in the M & M section. If some of this information were put on the graph itself, it would unacceptably clutter it up. The legend also points out a particular feature in the graph that helps the reader to understand it better while checking it.

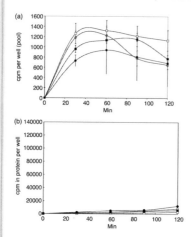

Fig. 5. Uptake and incorporation of ^{14}C-argininosuccinate into the acid-soluble and acid-insoluble compartments of cells as in Fig. 4(a) Uptake into the acid-soluble pool in uninhibited CHO cells ($\square$) and into cycloheximide-inhibited CHO cells ($\bigcirc$). The corresponding data for A375 cells is given by filled symbols ($\blacksquare$ and $\bullet$, respectively). Data are the means of 6 samples, with the error bars showing ±one SD. The data was highly variable and may have related to the washing procedure of the cells, since the uptake and release of argininosuccinate was very rapid across the membrane. (b) Corresponding data on the uptake into the acid-isoluble pools. In both cases the data for the cycloheximide-treated cultures and the controls were indistinguishable, except perhaps for a slight increase in A375 cells not given cycloheximide where some AS conversion to arginine was just detectable ($\bullet$). However, to all intents and purposes, no significant incorporation into protein occurred in the other cases. Data are the means of six samples, with the error bars showing ±one SD (error bars fall within the symbols and have been left out to avoid confusion on the baseline of the plot).

present in the medium, eliminating any suggestions that the uptake in label was due to contaminating arginine in the preparation.

4. Discussion

Cultured cells exhibit a wide range of ability to handle citrulline and AS when arginine is missing.

Data in Tables 1–3 provide a valuable database on the integrity of the ornithine to arginine sector of the urea cycle in many cells types, showing clearly where it is deficient. Repeated analyses of many different 'sub-strains' of the same line, especially in Table 3, show most convincingly that cell lines 'breed true' over many years, retaining the same typing. Only with this knowledge will data on ASS activity and mRNA transcripts be meaningful in terms of the innate machinery that different cell lines retain to carry out the conversion of citrulline through AS to arginine, which is patently our next task, although others [2,21,22] claim that there is this persistent deficiency in melanomas and hepatocellular carcinomas. Possibly the best example of this remarkable stability of expression of the ASS/ASL pathway in such classic tumour line is in the B16 F10 mouse melanoma (Table 3), which has been extensively cultured over many years and is commonly used for in vivo experiments. The sub-lines possess a wide range of different behaviours, including large differences in metastatic potential. Given that cells retaining the integrity of the appropriate part of the urea cycle could at any time lose it by some gene deletion event, the fact that the F10 line has been grown in a wide range of media and conditions, but still retains this, as well as all the other substrains, indicates its importance where cells grow almost as fast on this substrate as on arginine. Eagle [23,24] had missed out citrulline in his early work on cell requirements for growth in culture, and later noted that it does substitute for arginine, although at least a two-fold higher concentration seemed to be about 'optimal'. Had he chosen CHO cells (Table 3), his edicts on culture requirements might have been quite different. The possibility of cells recovering the urea cycle activity after having lost it must be remote, which is why in contrast we would not have expect much variation in the CHO cell line.

Although some of the experiments reported in Figs. 2–5 seem obvious to the point of triviality, they were included because too often assumptions have been made in other studies, e.g. that small substrate molecules of metabolism cross cell membranes without difficulty. Unless we know this for certain, some of the conclusions of others are, but little more than inferences. We now know for certain that AS, like citrulline, has no problem gaining access to the cell

The Results section comes to a close in the left-hand column here. The two graphs at the top are intentionally labelled 'a' and 'b' so the reader will readily notice the comparison. Since the x axis (time in min) is the same, it is clear that the samples in Fig. 5b showed far less uptake than those in Fig. 5a, but the y axis had to been altered by two orders of magnitude so that the curves can be seen rising very slowly above the baseline as time passes in Fig. 5b, in sharp contrast to Fig. 5a.

The Discussion starts here at the bottom of the left-hand column. It is important that data/results are not reiterated at the start of this section. This Discussion opens with a general comment on the diverse abilities of cells to handle several amino acids. The rest of the section seeks to interpret the new data for the reader, to bring the separate aspects of the study together, and to consider them in relation to previous publications on the subject from the authors' laboratory and from other groups worldwide. It is important here not only to point out the novelty of the present findings and how they contribute to knowledge, but to decide whether they support or refute the hypothesis intimated in the Introduction. Science progresses by consensus, which is why data have to be evaluated in this way.

150   *D.N. Wheatley et al. / Cancer Letters 227 (2005) 141–152*

internum (Fig. 5), even in CHO cells. Indeed, both these compounds are taken up faster than a dibasic molecule, viz. arginine.

Apart from the few cell lines (A375, MEWO and Saos-2) that handle AS rather poorly as a growth substrate, no Type A cells can utilise AS even at high concentrations, and not because of problems of access.

Our data also confirms that Type C cells have the ability to utilise citrulline, but not AS [16]. Put another way, if the cell can utilise citrulline, it cannot use argininosuccinate. If it uses AS, then so far it would appear to date, it cannot utilise citrulline. This firmly establishes that the citrulline→arginine pathway is very tightly coupled [24,25]. It follows that when this pathway is disrupted through loss of ASS, ASL no longer receives a direct feed of nascent AS, and only a few cell types seem to be able to utilise rather inefficiently exogenous AS. Since Type B cells cannot use citrulline, the Type C (AS+) categorisation that we devised and set out in previous report [16] can now in all probability be abandoned because it seems most unlikely that it will ever have a representative. This is a very interesting and important observation, since it appears that all cells retain both the ASS and ASL genes ([25]; see also [26]), and what we detect is only the capacity of the cells to express their mRNA products and/or the enzymatic activity of their protein products. Where activity seems to be completely 'lost', we must in fact conclude that it must be highly repressed. To date, no one has found a situation where ASL enzyme activity has been completely lost. Thus loading the medium with ever increasing levels of exogenous AS for Type A cells (e.g. CHO) does no good, even though they retain the ASL gene.

Our argininosuccinate preparation was undoubtedly of low purity, and any arginine in it would have been capable of providing nourishment if present at even a few percent in the AS. Several factors indicate that this is not the case. First, the arginine would have been used up pre-emptively, such that a burst of proliferation would have been seen and then regress as it was exhausted. Any kinetics with AS indicate a slow rise in rate, commensurate with an entirely different mechanism that cannot be related to arginine. The possibility has been experimentally eliminated by showing that AS gives the same kinetics in A375 and MEWO in

the presence of arginase, which would eliminate any arginine released. The fact that these cells types cannot handle citrulline does not eliminate that the possibility of citrulline being a possible contaminant—a complication that has to be borne in mind in other studies, but is irrelevant in this context.

The Tables would become meaningless if the same cells had fluctuated between the different Types through sojourns in different media, conditions, handling, etc. In some cells, adaptation has been reported first by Jacoby [27] and later by Sugimara et al. [28], who found that lymphoblastic cells have a 150–200 fold rise in ASS activity, admittedly from extraordinarily low background levels of the normal lymphocyte, when cultured long term in medium in which citrulline has been substituted for arginine. This is also claimed to be the mechanism by which resistance is thought to develop to the toxic arginine analogue, L-canavanine [29]. After extensive work, we have found no evidence of this adaptation has been carried out, in addition to that reported in Fig. 3, in several cell lines (to be reported). Long-term exposure to citrulline has not in our hands induced levels of ASS that has changed a cell that could not previously use citrulline into one that can now survive and grow on it. Tanaka et al. [30] noted a similar induction, which therefore is relevant to the development of resistance to arginine deprivation where citrulline is available. A lot more needs to be learned about the regulation of ASS in both normal and cancer cells since the data are not compatible with our findings that indicate the cells do not adapt to the circumstances when obliged to use either citrulline or argininosuccinate in the absence of arginine, and this accords with the findings of Schimke [31].

Acknowledgements

We wish to thank many people for generously providing us sample of cells from their stocks to make many of the comparisons herein reported. They are listed below, and we apologise for any omissions from the list.

Our grateful thanks also go to Dr Oliver Musgrave (Department of Chemistry, University of Aberdeen)

The Discussion continues here. This section is not particularly long (Chapter 4 discusses the problems of overworked Discussions). On reflection it might have been better to break it into subsections. Scientific papers are seldom perfect in construction; remember a scientific paper is a 'state-of-the-art' document. In all probability it will in time be superseded by new findings, so its content should not be seen as set in stone. This section should not start by repeating the main results; it is best not to include more than the most salient points (4–5) or speculation that goes way beyond the scope of the paper. The Discussion ends with an overall summing-up, which suffices for this journal. In other journals a small section with the subheading Conclusions is appended after the Discussion.

The Acknowledgements section starts here at the bottom of the right-hand column. It first acknowledges in general the many contributors of the cell lines we used, but their names are at the end of this section. Thanks are given then to people who provided other materials and advice, before thanking colleagues for technical support. The two paragraphs that appear next, just before the Reference section, are essential.

D.N. Wheatley et al. / Cancer Letters 227 (2005) 141–152   151

for kindly preparing ^{14}C-argininosuccinate. Liu Yunli and Nicola McKernan kindly provided technical help. This work was supported by grants from the Synos Fondation (Switzerland), Grampian University Hospital Trust Endowments Fund, and the 'Friends of Anchor' (Balmoral) Cancer Group, Scotland, to whom we are most grateful.

Cell line suppliers have included Dr Paul Winyard (London); Dr Anthony Chang (Hong Kong); Dr Bon Hong Min (Seoul); Dr Isabel Smith-Zubiaga (Bilboa); Dr Sam Bowser (Albany); Dr Stephen Cooper (Ann Arbor); Dr Sam Crouch (Dundee); Dr Sally Wheatley (Sussex); Dr Kazuto Kajiwara (Madison); Dr Tim Illidge (Southampton); and others.

References

[1] H. Gong, F. Zölzer, G. von Recklinghausen, J. Rossler, S. Breit, W. Havers, et al., Arginine deiminase inhibits cell proliferation by arresting cell cycle and inducing apoptosis, Biochem. Biophys. Res. Commun. 261 (1999) 10–14.

[2] C.M. Ensor, F.W. Holtsberg, J.S. Bomalaski, M.A. Clark, Pegylated arginine deiminase (ADI-SS-PEG[20,000mw]) inhibits human melanomas and hepato-cellular carcinomas in vitro and in vivo, Cancer Res. 62 (2002) 5443–5450.

[3] I.-S. Park, S.-W. Kang, Y.-J. Shin, K.-Y. Chae, M.-O. Park, M.-Y. Kim, et al., Arginine deiminase as an inhibitor of angiogenesis and tumor growth, Br. J. Cancer 88 (2003) 679–687.

[4] E.-J. Noh, S.-W. Kang, Y.-J. Shin, S.-H. Choi, C.-G. Kim, I.-S. Park, et al., Arginine deiminase enhances dexamethasone-induced cytotoxicity in human T-lymphoblastic leukemia CCRF-CEM cells, Int. J. Cancer 112 (3) (2004) 502–508 (ahead of hardcopy).

[5] J. van Rijn, J. van den Berg, T. Teerlink, F.A. Kruyt, D.S. Schor, A.C. Renardel deLavalette, et al., Changes in the ornithine cycle following ionising radiations cause a cytotoxic conditioning of the culture medium of H35 hepatoma cells, Br. J. Cancer 88 (2003) 447–454.

[6] F. Izzo, P. Marra, G. Beneduce, G. Castello, P. Vallone, V. De Rosa, et al., Pegylated arginine deiminase treatment of patients with unresectable hepatocellular carcinoma: results from phaseI/II studies, J. Clin. Oncol. 22 (2004) 1815–1822.

[7] J.S. Bomalaski, J.L. Ivett, M. Vegarra, F.W. Holtsberg, C.M. Ensor, M.A. Clark, Comparative toxicity of arginine deiminase formulated with PEG poly(ethylene glycol) of 5000 and 20,000 molecular weight and the effects of arginine deprivation in mice and rats, Preclinica 1 (2003) 284–293.

[8] S.A. Curley, J.S. Bomalaski, C.M. Ensor, F.W. Holtsberg, M.A. Clark, Regression of hepatocellular cancer in a patient treated with arginine deiminase, Hepato-Gastroenterology 50 (2003) 1214–1216.

[9] S.J. Bach, D. Swaine, The effect of arginase on the retardation of tumour growth, Br. J. Cancer 19 (1965) 379–386.

[10] J.M. Storr, A.F. Burton, The effects of arginine deficiency on lymphoma cells, Br. J. Cancer 30 (1974) 50–59.

[11] L. Scott, J. Lamb, S. Smith, D.N. Wheatley, Single amino acid (arginine) deprivation: rapid and selective death of cultured transformed and malignant cells, Br. J. Cancer 83 (2000) 800–810.

[12] D.N. Wheatley, E. Campbell, Arginine catabolism, liver extracts and cancer, Pathol. Oncol. Res. 8 (2003) 18–25.

[13] R. Philip, E. Campbell, D.N. Wheatley, Arginine deprivation, growth inhibition and tumour cell death: 2 enzymatic degradation of arginine in normal and malignant cell cultures, Br. J. Cancer 88 (2003) 613–623.

[14] D.N. Wheatley, Controlling cancer by restricting arginine availability—arginine-catabolizing enzymes as anti-cancer agents, Anti-Cancer Drugs 15 (2004) 825–833.

[15] J. van Rijn, B.J. van den Beerg, R.G. Schipper, S. de Jong, V. Cuijpers, A.A. Verhofstad, T. Teerlink, Induction of hyperammonia in irradiated hepatoma cells: a recapitulation and possible explanation of the phenomenon, Br. J. Cancer 91 (2004) 150–152.

[16] D.N. Wheatley, E. Campbell, Arginine deprivation and tumour cell death: 3 deficient utilisation of citrulline by malignant cells, Br. J. Cancer 89 (2003) 573–576.

[17] H. Takaku, S. Misawa, H. Hayashi, K. Miyazaki, Chemical modification by polyethylene glycol of the anti-tumor enzyme arginine deiminase from Mycoplasma arginini, Jpn. J. Cancer Res. 84 (1993) 1195–2000.

[18] K. Sugimura, T. Kimura, H. Arakawa, T. Ohno, Y. Wada, Y. Kimura, et al., Elevated argininosuccinate synthetase activity in adult T leukaemia cell lines, Leuk. Res. 14 (1990) 931–934.

[19] E. Bishop, G.D. Bell, S. Bloor, J. Broom, N.F.K. Hendry, D.N. Wheatley, An in vitro model for angiogenesis: basic features, Angiogenesis 3 (2000) 335–344.

[20] C. Rodriguez-Ayerbe, I. Smith-Zubiaga, Effect of serum withdrawal on the proliferation of B16F10 melanoma cells, Cell Biol. Int. 24 (2000) 279–283.

[21] K. Sugimura, T. Ohno, T. Kusuyama, I. Azuma, High sensitivity of human melanoma cell lines to the growth inhibitory activity of mycoplasmal arginine deiminase in vitro, Melanoma Res. 2 (1992) 191–196.

[22] B.J. Dillon, V.G. Pieto, S.A. Curley, C.M. Ensor, F.W. Holtsberg, J.S. Bomalaski, M.A. Clark, Incidence and distribution of argininosuccinate synthetase deficiency in human cancer: method of identifying cancers sensitive to arginine deprivation, Cancer 100 (2004) 826–833.

[23] H. Eagle, Nutrition needs of mammalian cells in tissue culture, Science 122 (1955) 501–505.

[24] H. Eagle, Amino acid metabolism in mammalian cell culture, Science 130 (1959) 432–437.

Funding: The third paragraph of Acknowledgements, at the top of the left-hand column here, provides important information about the funding of the research work. When there was only one major funding body (e.g. the Medical Research Council), the grant number would usually be given. In this case, several sources provided support for the work described in the paper. If any of these different funding sources had related to particular authors, the initials of those authors would have been included in parentheses after the relevant grant.

The Reference section (or Bibliography) follows the Acknowledgements. With the Vancouver system – as used in this demonstration paper – the references are numbered in the order in which they appear in the text (e.g. [1], [2], etc.), so there is no easy (alphabetical) way of checking that a citation from a particular author has been included, except by going down the whole list. What the Vancouver system does is reduce the length of the text citations. The alternative (Harvard, author–date) format – e.g. (Bouvier and Jones, 2008) – is convenient for noting in the text who wrote a cited paper, but it uses a little more space.

The journal publisher decides which system to use and authors have to comply. The way in which each citation is presented in the reference list must also conform to the style required by the journal; modern computer software for reference databases can handle this for you. (In this paper, there is another page of references, but this has been omitted because it is unnecessary to continue any further for purposes of illustration.)

Now that we have finished the notes accompanying the paper, you have seen a published example of a primary research paper. A quick browse through papers that can be accessed easily online will show you the diversity of the scientific literature in the way papers are presented by different journals. However, although you will find these variations in presentation, there is overall remarkable similarity and uniformity in the format of papers, section by section. In older journals, for example, very often the abstract would be found just before the reference list, which defeated the purpose of having it up front where it can easily be read to see if the rest of the article is worth reading. Often they also provided abstracts in English and the language of the country of origin, e.g. English then Russian (or vice versa).

What has to be re-emphasized is that modern-day publishing puts the *onus on the author* to comply with the demands of the journal to which the paper is going to be submitted. All journals have 'Instructions to Authors' sections on their websites. It is imperative that you take heed of those instructions if you wish to have a smooth path to publishing good papers. The following chapters will give detailed instructions of the drafting of papers right through to the publishing procedure.

Chapter 3 will show you how to prepare the **Results** section of a paper. The reasons for this being usually the first section to draft were discussed in Chapter 1.

3 Results
Presenting Your Findings

A research article is the *final product* of an investigation. It is a report that tells the world what you have done and found. On this account, it must be presented in the best possible way so that its message is to the point, clear and succinct.

3.1 Starting Out: Getting the Data Together

In Chapter 1, the Results section was described as the 'pivot' of a paper. It sets out the information you wish to communicate. I strongly recommend that you draft this section of your manuscript first, although one most important issue must be kept in mind, collating the relevant findings!

Suppose you have been working on your research topic for 6 months or a year. By now you probably have substantial findings to share. One day you are asked to give a seminar to the department and probably you will find the best way to proceed is as follows:

- If you have not already done so, gather the data from your notebooks together accurately and presentably in the form of graphs, images, tables and so on.
- Put these items into PowerPoint.
- Review all the PowerPoint slides and put them into a logical sequence that best conveys your findings and sustains your arguments clearly to the audience, placing the emphasis on the most important.
- Decide what you are going to say about the data. It is usually this way round – data first, comments second – not the reverse.

Although giving a seminar is not the same as submitting a paper for publication, there are similarities. The seminar exercise makes you gather the essential data together to make a coherent 'story'. These data form the framework on which your presentation must hang. The same principles and procedures apply to writing a paper, though in a paper you do not explain things as you go along, as you do in a seminar. In the Results section of the paper you need to set out your data (graphs, images, figures, tables) as succinctly and clearly as possible. What prompted you to research the topic and how your findings relate to other peoples' data will be dealt with in other sections of the paper and discussed in subsequent chapters.

Results – the first job

Find the relevant figures, tables, graphs and others (the raw data) that you need to communicate, prepare them in a good format and review them so that the crucial ones can be identified and presented first.

3.2   Arranging the Data

Your data need to be arranged in the order that makes the points necessary to sustain (or refute) the **hypothesis.** Always keep the hypothesis in mind; not everything you have done or recorded will necessarily be appropriate for your paper and many of your experiments will have little or no relevance to your arguments. The selection process is critical because people (e.g. you and your co-authors) may disagree about what is and is not worth including (see Sections 3.3 and 3.4). For the sake of succinctness, it is important that data 'peripheral' to the main theme of the paper are identified and left aside until after the Results section has been constructed. These might then become supplementary data, which can be reviewed later to see whether any of them merit inclusion (e.g. do they provide extra controls that help to verify a major finding? See Section 3.8).

A scientific paper is not a diary; a diary is a chronological record. Since the early twentieth century, authors of scientific papers have not been obliged to record findings in the sequence in which they conducted their experiments. Sometimes the very last set of experiments you do proves crucial and needs to be communicated first; these findings can quite unexpectedly often make all your previous work fall neatly into place. Thus, it is the *logic* that is important, not the *chronology*. Nevertheless, there is a strong tendency amongst researchers, and not only novices, to sustain the 'diary' tradition. Only an exceptionally gifted (or lucky) researcher might go from one set of experiments to the next so that the logical sequence of study coincides precisely with its chronology.

Chronological reporting in old journals

If you look at papers published in the late nineteenth and early twentieth centuries, you will be surprised to find they were often written in something akin to diary form. They were like tidy laboratory notebooks, mentioning everything that was done and what was found, usually in strict chronological order. Pasteur's papers are excellent examples.

3.3   The Selection Process: Further Considerations

Most papers have one major message to communicate to the reader, so one experiment will probably be of particular importance and everything else that is included will depend on it, and the other data should lend good support (assuming your hypothesis is correct!). Get your co-authors to agree about this. Put it up front and state it as boldly as you can with as a little 'dressing up' (fine detail) as possible.

The ancillary experiments supporting your major claim may be related to each other in an obvious sequence, but sometimes they impact on the main finding from quite different directions, e.g. kinetic data are unlikely to relate directly to ultrastructure. It is up to you to decide the best sequence of presentation. Your PowerPoint slides when set out in review mode will help you with the selection process, but remember to set aside the more trivial data in case you have to come back to them (see above). From about 1930 until 1960, many published papers ran to 40–50 pages, perhaps representing several years of work. Today the emphasis has changed and papers are usually less like theses. Do not overcrowd your Results section with an excess of figures, tables and images.

Long and short papers

Take a look at the very first paper ever published in *Experimental Cell Research*. The author was a certain Francis Crick! It is about 50 pages long, and its hypothesis was flawed. In contrast, his seminal paper with James Watson in *Nature* on the structure of DNA ran to just one and a half pages. The latter paper is a classic while the former has rarely been cited.

Current practice favours publishing a series of papers rather than one long diatribe. There are many reasons for this, though not all scientists would agree. For instance, the number of papers a research group has published is widely regarded as a measure of the authors' achievements and progress. Also, many readers (and journal editors) dislike excessively long papers. Never forget, however, it is quality that should be assessed by others, not quantity.

3.4   Positive, Negative and Neutral Evidence in the Selection Process

The most difficult part of the selection process is to decide not just what to *include* from your coverage of the figures, graphs, etc., but what to *exclude* after the pivotal information has been identified. There is a tendency to include only the data that sustain your hypothesis. Many of us choose to ignore the 'nasty little fact' that seems inconsistent with the rest of the evidence. That nasty little fact means your hypothesis and the reasoning behind it could need (possibly serious) modification, even complete rejection. This is negative evidence, and few journals are prepared to publish papers presenting purely negative evidence refuting someone else's hypothesis, unless the topic is of major importance, e.g. nuclear fusion in water at room temperature. However, negative evidence can be published when it is accompanied by positive evidence for an alternative hypothesis.

The question remains – is it ethical to omit such evidence from your own paper? The best policy is to include it, pointing out that it does not fit with the rest of the data, so your hypothesis could need amendment. If you cannot explain the problem, it can be discussed in the following section so that readers can make up their own minds about it and consider whether a new course of action (further experimentation) is needed to solve it. It gives food for thought and serves to promote more research by yourselves and others, which might lead to an entirely different hypothesis and perhaps open up new horizons in the research field. Your readers often see something in these seemingly difficult findings that you have missed, largely because they will be seeing the problem from different angles than yourself. A rogue point might well be explained by another person.

Unfortunately, present-day science is largely driven by funding and the need to publish positive findings. In addition to the positive findings in the paper you are drafting, you will probably have done many experiments that proved nothing, i.e. they neither refute nor support your hypothesis. We may refer to this as *neutral* evidence. Perhaps you repeated an experiment five times and only twice did you obtain a positive response, the other three trials showing no effect. What should be done in such cases? Everyone's instinct is to refer only to the two trials that 'worked'. Again, however, honesty is the best policy. The two positive results demonstrate that the phenomenon can happen, but not every time, which should make you seek a rational explanation for the variability. Maybe the conditions were not always precisely the same, and with careful scrutiny you may be able to find what was responsible for the apparent inconsistency. This could prove to be valuable information.

Popper's dictum

A leading philosopher of the twentieth century, Karl Popper, believed – as we all ought to know – that science progresses by consensus, and that sustainable hypotheses can be built up further as more positive evidence becomes available, culminating in them becoming tenable theories or laws. However, his key message was that we should also design our experiments to seek evidence to *refute* or *falsify* our hypotheses. If it withstands such thorough scrutiny for long enough, a hypothesis increasingly becomes part of our received wisdom. Sadly, few investigators these days design experiments that deliberately challenge and try to refute rather than uphold their cherished hypotheses. There is more to tell on this matter under 'The Hypothesis' in Appendix 5.1.

3.5   Adding the Text

So far we have said nothing about adding text to the Results section of your paper; we have 'written' nothing. Remember the exercise of giving a seminar, in which you probably presented your data and talked about each slide as it was shown. Writing the text can be done in the same way, around your selected figures and tables. But remember what was said above: refrain from entering into discussion at this time, since your next section (Discussion) serves that purpose.

3.6   A Note on the Use of the Personal Pronouns, 'I' and 'We', in Writing a Scientific Paper

As an editor, I find repetition of the phrase 'we did this. . . and then we did this. . .' annoying. The reason for doing an experiment or a set of experiments can usually be inferred from the fact that the authors are making a statement about it! So these phrases are redundant and editors are relieved when they are omitted. Try the exercise yourself if you have repeatedly used 'we' in your draft paper. You will find it much simpler and clearer to say what was done/found than to tell your reader that you did this first, and then something else second, and so on. Remember what was said above; chronological reporting is not necessary and is often inappropriate. Too many papers are written as if the work was commonplace or simple, carried out in an obvious sequence of experiments that proved each issue as you proceeded. This is more akin to a technical exercise than real science.

About 50 years ago, the use of 'I' or 'we' in a scientific document was frowned upon. Most editors and journals demanded that the impersonal, *passive,* voice be used. Instead of 'We investigated the effect of elevated glucose levels by measuring...' one had to write 'The effect of elevated glucose levels was measured by....'. There is no reason nowadays to use the passive voice exclusively, but a Results section (or any other section of a paper) in which the active voice is persistently used will grate on the reader. My advice is to moderate the use of personal pronouns so that fewer than half your sentences start with them.

Writing 'we did this... then we did this...' is common practice today. I acknowledge that it is difficult for writers whose native language is not English to use the passive voice efficiently and naturally; it is likely to be unfamiliar. For this reason, it is sometimes best to seek help from professionals. However, it is a useful thing for everyone to learn to use properly. We will deal with this again in later chapters where use of the English language will be discussed. For the time being, an example will help to clarify what I wish to indicate about 'acceptable style':

> We measured the rate of uptake of x as a function of y, using the method of Barnsworth (ref.). It was found that x increased in direct proportion to y when the temperature was kept at 32 °C.

is better than:

> We measured the rate of uptake of x as a function of y, using the method of Barnsworth (ref.). We found that x increased in direct proportion to y when the temperature was kept at 32 °C.

Note that the second sentence in the first version starts in the passive voice ('It was found that...'), but there is an even better way of writing that sentence. It can be inferred from the first sentence that something had been measured. This makes the expression 'It was found that...' redundant. The sentence can simply begin 'x increased in direct....'. The two sentences are so interconnected that if you were to read them aloud it would be like putting a semi-colon between them. The writer aiming for real succinctness without losing any vital information could go further and write the following:

> The rate of uptake of x as a function of y, measured by Barnsworth's method (ref.), was directly proportional at a constant 32 °C.

The personal pronouns have vanished altogether and the resulting short sentence contains all the evidence that you want to convey.

3.7   Checking the Logic and Order

To you and perhaps your co-authors the order of presentation you have chosen may seem optimal. If you are not the principal investigator (PI), it is advisable to seek his or her opinion at this stage. The PI will be familiar with what you have been researching; they may advise you that, seen from a broader perspective, there could be a better order. If you are a PI, I would recommend seeking advice from other colleagues (giving a seminar on the paper can serve this purpose). If those colleagues also review your supplementary material, they may help to identify experiments that ought to be transferred to the main body of the paper, or vice versa. Another relevant point is the need to repeat experiments several times (usually a minimum of three) in order to ensure the results are reproducible. If each experiment produces closely similar data, the results can be pooled, which will probably boost their statistical significance. The alternative is to display the data from one of the experiments as a representative result.

3.8   Controls: Have You Included all the Essential Ones?

I have stressed the importance of controls. I often see papers from which controls have been omitted, weakening their credibility. Many papers purport to show some effect in the experimental group that differs significantly from the control group; however, two groups are sometimes not enough to establish the difference between them beyond all reasonable doubt. This is seldom obvious to the less experienced researcher. An example will help to clarify the point.

Let us return to the previous discussion about the rate of uptake of x as a function of y. Suppose rats were injected with different doses of substance y, and the uptake of x (e.g. glucose) through the gut mucosa was measured. The obvious control is to have rats injected with only the vehicle in which substance x was dissolved in the experimental group. However, that control alone will not suffice. The rats could have been increasingly stressed by the injections, so the rate of uptake of x could have been affected by the handling and injection procedures. To eliminate this possibility you need *two* further control groups, the first comprising rats that were handled but not injected, and the second comprising rats that were not handled at all. There may be no differences in the uptake of x among these three control groups, but that needs to be demonstrated in early experiments with such a model. Provided the 'extra' controls show no obvious (statistical) differences, a brief mention of them in the Results (without the actual data) could suffice. The confidence that statistical differences in the uptake of x between y values are 'real' is boosted if you have done all three controls, which might then be omitted from later experiments in the series.

3.9   Statistics

The results section should deal (along with the figures and tables) with the statistical significance of differences between experimental and control groups, usually expressed as P values (P for probability – upper or lower case can be used, but be consistent). In biological experiments, a probability of 0.05 or less is generally accepted (i.e. one trial not 'working' out of 20), although different criteria can be set. If $P = 0.1$ or thereabouts it is usually considered to indicate a trend rather than a significant difference. These ideas are difficult for those who have not undertaken a course in statistics, and even if you have taken a basic course, nevertheless I would encourage you to seek advice from a professional statistician. Appropriate statistical tests must be used to compare data. This is a specialized field in which younger researchers need good advice. However, while acknowledging the obvious importance of statistics, I do not intend to discuss them further in this book. Most authors use software that usually provides appropriate tests that will indicate the significance of results, but not everyone knows exactly which test is best to apply, and therefore it is always worth consulting experts. They must, however, be made fully aware of the basic problem that you are trying to solve and its particular context. It is also advisable to know yourself what statistical manipulations are doing to your data. Because of the availability of sophisticated software, unfortunately the vast majority of authors do not understand what statistics measure and how calculations are done.

3.10   Summary: the Major Points of a Typical Results Section

- State your main finding at the start of the Results.
- Indicate straightaway what is new (i.e. original) about it.
- If it in some way corroborates previous work by another group, you might want to say so briefly here, though you will expand on the point in the Discussion.
- Provide the rest of your relevant experimental data in a logical sequence, without going into further detail at this stage.
- The data will include your figures, graphs, images and tables, with legends that give short descriptions of what was done, but normally leaving technical details to the Materials and Methods section.
- Statistical information should be included in the graphs and figures. You do not need to provide all this information in the text – you can simply refer to the appropriate graph for the reader to see the P values or the regression line through a set of data points.
- Refrain from including unnecessary (insignificant) experimental data.
- Include 'negative' evidence that for some reason seems inconsistent with most of the other findings, but you do not discuss it until the next section.
- Figure legends, tables, micrographs and others are normally appended to the manuscript after the Reference section. The publisher will find suitable places to insert them, although you might wish to indicate where you would like them to be placed relative to the text (e.g. 'Figure 1 here'), either in a space in the text or as a marginal note.

The major findings can be mentioned in no more than two sentences at the end of the Results section, in preparation for the opening of the Discussion. The results should not be reiterated at the start of the Discussion section. The Results section, if set out as indicated, might be the shortest part of your paper. I have seen papers in which it occupied only about a third of a page, while the Introduction and Discussion sections each occupied several pages.

4 Discussion
The Place to Argue Your Case

A research article is the *final product* of an investigation. It is a report that tells the world what you have done and found. On this account, it must be presented in the best possible way so that its message is to the point, clear and succinct.

4.1 The Basis of a Good Discussion

One of the most important things to remember when you start the Discussion section is *not to reiterate or summarize your results*. To do so is becoming a habit among authors. It should be avoided at all costs. It is a major contributor to the repetition and redundancy that are all too common in present-day papers.

If your findings confirm your hypothesis beyond reasonable doubt, you need only a short Discussion to set your new findings in a broader context and to tie up loose ends left by previous research. No more should be needed (remember Crick and Watson's *Nature* paper, mentioned in Chapter 3). However, few papers 'confirm' a hypothesis so comprehensively. Doubts will remain in the overwhelming majority of cases and your Discussion should address them – as well as relating your results to the wider context. And if your work has refuted the underlying hypothesis and perhaps led to a new one, you will have plenty of explaining to do.

That sums up the content of a Discussion section: the findings you reported in the Results section should be explained, both in themselves and in relation to other published reports. It is usually best to cite the papers that led you to your hypothesis, experimental design and results, since their authors may have been the proverbial 'giants on whose shoulders you stood'. No piece of research is ever done in isolation. However, the Discussion is not the section of the paper in which earlier work per se should be reviewed; more about this later.

4.2   State-of-the-Art: Never Absolute Truth

Remember that scientific papers report the *state of the art*; seldom will they approach the 'absolute truth' (an immutable fact) or even be particularly durable. Facts are also dependent on circumstances and conditions, e.g. if we state that water is a liquid, this is not true below 0 °C or above 100 °C. The findings may therefore be valid, but perhaps only in a narrow context. This must be kept in mind at all times when preparing a Discussion. Your results can range from very convincing to only mildly convincing, perhaps because your hypothesis is difficult to address experimentally and thorough testing would take years or be too expensive (see Chapter 3). Almost every finding will require more work – research is never-ending. The weaker the supporting evidence, the more you will need to argue your case, which will lengthen the Discussion. Note that the stronger the evidence, the shorter the Discussion.

It is good policy to apply Karl Popper's prescription: try to *refute* your hypothesis. If it withstands strong challenges and proves difficult to refute, it will carry more conviction, at least for the time being. This brings us to an important point about submission – too often, the work described in a manuscript is insufficient to justify the claim that a hypothesis has been adequately tested. The researchers are overly eager to publish before the evidence for (or against) the hypothesis is convincing. The Discussion is accordingly thin and weak, indicating that it would have been much better to defer submission until the case, for or against, was more convincing. One of the hardest tasks, as indicated in the previous chapter, is to decide when to write a paper.

Science moves forward by consensus and, in practice, most of us design experiments to support and corroborate our pet hypothesis, not to falsify it. However, you should be prepared to discuss results that are incompatible with your hypothesis, and perhaps with received wisdom. Focusing on the weaknesses of an idea often leads to better ideas, i.e. to an improved or refined hypothesis.

4.3   Putting Your Findings into Context

If your findings agree with existing papers, is there enough novelty in your work to merit publication? The purpose of a paper is to impart worthwhile new evidence or new ideas. Or is your evidence only indirectly, tangentially or incidentally relevant to a major problem in the field? Even if your data are good, do they contribute sufficiently to the advancement of knowledge?

If you are satisfied on these points, you have to justify your position. *That is the main role of the Discussion.* Ask yourself the following questions:

- Are your findings more credible than those of others? If so, why?
- Did you take a different approach from previous authors addressing the problem?
- Were your experiments better controlled or designed, yielding different or more accurate results?
- Did you advance knowledge in your subject area/discipline significantly enough to generate an improved or a new hypothesis?

Imagine you are an independent assessor of your own results. Are you satisfied with them, or do they have shortcomings? If you cannot see any shortcomings, discuss your findings with colleagues over coffee rather than staying in splendid isolation in your office. Even if you can defend the findings vigorously, others might view them sceptically, identifying problems that are likely to be the basis of challenges by other researchers in the field. The more such issues are raised, the longer your Discussion section will tend to grow.

However, referees, editors and readers dislike interminable discussions. Keep the virtue of succinctness in mind and focus on what is important. That will guide you to writing a clearer and shorter Discussion, and hence, probably, to more clear-cut conclusions.

4.4   Knowing Where to Start

Writing a Discussion is not easy, even for experienced scientists, so let us consider how best to organize it. It is usually the section on which most time has to be spent; do not let yourself feel frustrated if progress is slow. A Discussion requires much forethought and organization.

Making a good start

An example of the start of a Discussion that should be avoided

Discussion

> We have investigated the way in which CYTp1 interacts with Apel2 in the mouse 3T3 fibroblasts system. Our main result shows that when both genes were fully expressed under normal conditions of cell culture, intracellular glucose levels rose from 12.3 µM to 21.9 µM.

The reader will know what has been investigated by the time the Discussion is reached; there is no need for an opening sentence that repeats this information! The second sentence reiterates details dealt with immediately above in the Results section. These days, this regrettable type of opening is common in the Discussion sections of published papers.

4.4.1   Important Tips on Getting Started

The Results section gave details of the experimental and/or theoretical findings. Let us assume you presented them clearly and unambiguously, so they should not be repeated at the start of the Discussion unless some of their implications for the hypothesis are not obvious. The main task at the start of the Discussion is to identify the issues that are the real substance of the paper, and the problems arising from interpretation of the data. List the issues in order of importance, and then deal with them in that order.

Select just four or five issues from the Results section that agree (or do not agree) with your hypothesis or with previously published data. Leave aside the findings that are corroborative but could not stand alone. (It often is the case that the much less important points that are left out have a habit of being understood *implicitly* when a well-argued case is made with the main points.)

This selection will bring the task of interpreting the essence of your research into better focus. This is particularly important where the context is difficult and/or the arguments for and against a hypothesis are diverse. However, the points should be easily recognizable. Most experienced researchers and writers know where to place the emphasis and how deeply they need to delve into further analysis to support a hypothesis.

4.4.2   Subheadings

It is often good policy to use the major points as subheadings in the Discussion. Experience is a great asset in this task, and it is worth persevering with exercises of this kind to improve your proficiency in writing. It cannot be repeated too often that *the Discussion should not reiterate the details presented in the Results*. To bring this point home, look at the next two boxes. Also, beware of using the Discussion as a repository for data, analyses and information that should be in the Results section.

Inadvisable repetition

The following is an example of an inadvisable repetition of results data in a Discussion:

A large decrease in the concentration of X from $340 \pm 23 \, \mu M$ in the control rats to $103 \pm 17 \, \mu M$ in those treated with Y was observed. This was seen as an important finding in this study.

A preferable wording in the Discussion section:

An important difference noted between control and treated rats was that the concentration of X dropped to about a third in the latter group.

The second statement starts by saying there is something of importance here. It then mentions a figure of 'about a third' of the concentration of X, obviating the use of 'exact figures', which would not be identical if the test were repeated several times. So the data have been simplified to relative values (a round figure), which is what the reader needs to know. He or she can go back to the Results section to check the actual figures if desired.

It is important to confirm the context in which your findings should be seen; after all, discoveries and novel observations can usually be attributed to insights arising from earlier work. But remember, some of this will have been covered in the Introduction, where the contextual background has been set out and explained, so you should check your opening remarks. The contents of the Introduction and the Discussion need to be balanced, each section complementing but not reiterating the other. Reiteration annoys referees and editors – and readers.

An inadequate Discussion

Problems commonly seen in an inadequate Discussion:

(1) It is poorly structured, and possibly fragmented;
(2) The text does not flow, is not coherent or logical;
(3) The context is ill-framed;
(4) It is too long;
(5) Too many tangential and irrelevant issues are discussed;
(6) Evidence that contradicts your hypothesis is ignored or summarily dismissed;
(7) The most salient elements are not drawn together to form a clear conclusion.

It is now obvious that the scheme I have outlined for constructing a scientific paper has generated a problem; you have not even started to write the Introduction! You will be doing so shortly, so this is a good time to decide how much background goes into the opening section and what can be integrated into the Discussion. At this stage, assume your Introduction will provide adequate background, and let the Discussion use this background without repeating it. Unnecessary repetition is easily spotted and it often ruins the chances of a doubtful paper being published. However, journals give no strict rules about this delicate balancing act because details differ from one paper to another, so you must decide how best to achieve it in your paper. My advice is to work hard at getting the balance right. Editors sometimes decide whether or not to accept a paper that is borderline on other criteria by assessing how well the material is organized and argued.

One way of keeping the reader involved and interested is to describe the background and immediate context loosely in the Introduction, allowing the Discussion to expand those aspects that most help you to interpret your findings. However, the opposite is sometimes better, i.e. to give a detailed explanation of the context in the Introduction, leaving more time in the Discussion to argue specific points. What you write in the Discussion influences the subsequent writing of the Introduction. In every paper you write, this balancing act requires thought before you commit anything to paper. Consider Mozart, he had all the ideas and information for a symphony all in readiness in his head – composing a score was then just a matter of writing it down.

4.5   The Strengths, and Weaknesses, of Your Experimental Evidence

The work of other groups on the phenomenon you have researched may provide perspectives that you had not fully appreciated before you wrote your paper. They may have brought the whole issue into better focus. In English, we use the phrase 'when the penny drops' to describe the coming together of apparently disparate observations into one obvious relationship. Everything suddenly makes sense. If you argue each point separately, you are likely to find yourself dealing only with specifics and overlooking the way your evidence fits the 'bigger picture'; the penny does not drop. There is an old saying that 'God is in the detail'. Nevertheless, without a grander scheme, the real significance of some small detail might not be appreciated. The details should be argued well, but we must also keep the bigger picture in mind, especially in the summing-up.

Results and Discussion

Many authors cannot, or do not want to, separate their results from a discussion of their significance. In some journals it is acceptable to have a single 'Results and Discussion' section, but if you want to communicate your data clearly without risking confusion, keep the two sections separate.

Some research findings can have an impact on fields of general concern, so they need to be interpreted and discussed in a particularly wide context.

Their importance might not be immediately apparent; a good example is work on the chemistry of CFCs, which widened into a debate about their effect on global warming. This is why a Discussion section can be so valuable. Also, the Discussion is often where the merit of an author can be judged! A paper can be considered outstanding because the results have been interpreted logically and cogently, so its hypothesis attains consensus more rapidly than that of a poorly argued paper with equally adequate results. The Discussion is the section that exposes authors who plagiarize other people's work because their arguments tend to be weak and inconsistent after the data had been presented boldly in the Results section.

4.6   Succinct Arguments and the Use of Inference

Excessively long arguments defending an average or a weak position should be avoided because they exasperate readers (including referees) as well as editors. Readers are happier if the weakness of the evidence is admitted and experiments that could improve the position are outlined.

Evidence presented in support of a hypothesis is always likely to have weaknesses, some serious and some less so. 'Weak points' do not have to be pieces of evidence that conflict with the hypothesis, but evidence that fails to substantiate the hypothesis as strongly as you would wish. In six months' time, you, or someone else, might substantiate the point but it just cannot be done with the information to hand. We seldom have all the evidence needed to uphold a hypothesis. There is no point in writing a long defence of a weak position; it is better to acknowledge its weakness and indicate how it can be strengthened.

The rule for achieving succinctness is simple: know exactly what the data tell you (and your readers), i.e. their interpretation. A violinist playing in front of a paying audience has prepared behind the scenes for weeks. The notes to be played for the benefit of the listener are all and only the ones written by the composer. In the same vein, you need to prepare so you can provide just the right words to get your message (the tune for the violinist) across. So the three things you have to consider are:

- your preparedness in knowing your material and its limitations intimately;

- the strength to resist bringing in anything superfluous; and
- the experience to be able to do this without confusing yourself or your reader.

These tasks might not seem 'simple', but the more you practise the better you will become (just like the violinist). And to push this analogy further, simple tunes are the most memorable. A succinct and well-argued Discussion that weighs the evidence as simply as possible is much more easily recalled and remembered and therefore exerts more influence on its readers than a long-winded and diffuse one.

There is one facet of succinctness worth mentioning from a purely literary viewpoint. So far, I have intentionally left aside the use of *inference*. Inference is often poorly used by scientists, and indeed by many present-day writers irrespective of discipline. The Japanese haiku illustrates what is important here. A haiku is a very formal poem; it has to be 17 syllables long in three measures (lines) of 5, 7 and 5. Imagine you want to describe the meeting of two old friends at sunset on a seaside promenade at dusk in just 17 syllables! The art of the haiku is to know what words are implied (would be made redundant) by what is written. What is said in two or three words can suggest as much as a long paragraph; their implication is clear. What your reader infers is as important as what you actually write.

You should not underestimate the intelligence of your reader. In a scientific paper, your readers often know as much about the subject as you do, perhaps more. Explaining the obvious wastes words and annoys the reader. To go back to one of my trivial examples, consider the sentence 'The sample was spun in an ultracentrifuge at 10,000 g for 30 min.' To spin something at this acceleration requires an ultracentrifuge, so it is implicit in the revised sentence 'The sample was spun at 10,000 g for 30 min.' Even in this short sentence, three words ('in an ultracentrifuge') are unnecessary.

4.7   The World of Ideas (and Dreams): the Problem with Speculation

Discussion of hypotheses arising from your results can be almost endless. This is the place to express your thoughts on the research that you have done along with the findings of others. The golden rule is never to speculate beyond what your results will allow. This is difficult to judge, even for the expert. So the practical guideline that accompanies the golden rule is: keep it circumspect, plausible and to the point.

You may want to let your imagination run riot at the end of the manuscript. There are at least three reasons why you should resist the temptation. First, if you indulge in wild speculations, your readers will probably not share them and might deem them crazy, especially if they appear in the wake of a thorough, careful and accurate piece of research that has been discussed logically within the appropriate framework. Second, you may have developed a new insight into the problem under investigation which, with a little further work, could lead you to new hypotheses, paving the way for future papers. Why give these revelations away when they are the seed-corn of new research? And third, speculations without evidence just add to the length of the paper, which in most cases is not worthwhile. Indeed, reviewers (and editors) often ask (or demand) that blatant speculation is removed before a paper can be deemed acceptable.

4.8   Is It Really Worth Publishing the Paper?

We now come to one of the most obvious – yet poorly appreciated – aspects of writing a good scientific paper. Suppose an author does not understand the full significance (or insignificance) of the findings, or has not created a coherent paper from the research investigation, but plods on and finally submits a manuscript. He or she is likely to be rebuffed by one journal after another, the referees recurrently indicating that the evidence is weak and/or poorly argued, or too few experiments have been done, or inadequate controls undermine the argument. In brief, the paper is seen as too preliminary, therefore taking stock while writing the Discussion is a worthwhile exercise. If things are not coming together and the evidence seems thin, you should be prepared to shelve the draft until your research has rectified the position and you are ready to say something significant. However much you may be disappointed after expending the effort to take a draft this far, it will not have been in vain. It will have been a valuable exercise, giving you greater insight into your own work. Rushing ahead to publish because of the pressure of pending grant applications or promotion is no justification. Unfortunately, this practice is prevalent because the competitive world of scientific research seems to demand it.

4.9   Reaching Conclusions

The final part of the Discussion should summarize what has been found in general terms and within its wider context. The main message is what needs to be emphasized here, in the most rounded terms (i.e. without detail), with minor points generally being left aside. The following, not uncommon, 'conclusions' are inexcusable: (i) grandiose claims that these research findings will help solve the problem of cancer, lead to a new understanding of evolution, and so on; and (ii) mentioning that further work is required to confirm the data and/or lead to new avenues of research. Of course more work would be needed if the results or conclusions are in any way dubious (as most are). Research is a never-ending task. Since science moves forward by consensus, others will also need to corroborate the findings. Also, any research activity can open up new avenues of opportunity for further work. There is no point in stating the obvious, so don't state it!

5 The Introduction
The First Major Section of a Paper

A research article is *the final product* of an investigation. It is a report that tells the world what you have done and found. On this account, it must be presented in the best possible way so that its message is to the point, clear and succinct.

5.1 A Brief Preface

In this section, which introduces the writing of an **Introduction** for a scientific article, the message is in the subheading: keep it brief. In a paper of 14–15 A4 sides of typing (1.5-line spaced), about 5000 words in total, it is normally best to limit the introduction to about one page. Think of one and a half pages as extravagant.

The main purposes of an Introduction

(1) To inform your readers of the *subject area* of your research, if possible, in one (or at most two) sentences.
(2) Following from this, to identify the *specific topic* of interest, and why it has been chosen.
(3) To outline the *specific problem* you wish to address in the paper.
(4) To set the context or background as succinctly as possible by mentioning key published papers – the most relevant and preferably recent ones – but *not* discussing them in depth at this stage.
(5) Finally, to state the hypothesis emerging from the foregoing exposition. Your paper will be about testing this hypothesis.

5.2   Informing Your Readers About the Subject Area

Remember that most primary research articles are intended for readers who are as expert in the subject as the authors. (This does not apply to review articles, which require a different approach.) Your readers will probably know the general context of your work as well as you do, sometimes better, so there is no need for a lengthy description or comprehensive review of the topic (although it is often helpful to cite a seminal review), or to discuss previous work in detail. If you identify a point of contention with previous findings then it might be useful to spell it out, but without detailing the underlying reasons. An Introduction should set the scene as succinctly as possible. It should not be too broad and general.

At triage (i.e. the first preliminary reading after submission), it can be seen that in cell biology and cancer journals, for example, many authors begin by stating what a problem cancer is, what a dreadful disease it is, how it affects people all over the world in different ways; then they go on to describe the incidence of some particular type, finally bewailing the fact that little progress has been made during the last 20–30 years. This type of very broad introduction is a waste of time in a primary research paper. All knowledgeable readers in the cancer field are aware of the general problem; indeed, the general public may know as much (point (1) in the box above). The authors then tell us that the particular cancer that they are going to discuss, say rhabdomyosarcoma, is especially intractable to treatment. Again, this information is unnecessary for your readers, most of whom will lock on to two discriminators in finding your paper if it is published (e.g. cancer + rhabdomyosarcoma). They are specialists and they know the problems associated with this tumour. Being succinct would be an opening such as: 'Breast cancer cells of rats differ in invasiveness compared to human beings.' Your next sentence(s) will clarify in which aspect of invasiveness you are interested, and this will lead on to a brief description of a particular problem that seems important and unresolved. On this basis, you have formulated a hypothesis that has been tested and on which you are now going to report.

A slightly longer exposition might be worthwhile if you have something to say that is potentially relevant to other cancers, since this could interest a wider medical and scientific readership. Nevertheless, brevity is essential, and the content, purport and focus of your paper must be kept firmly in mind before starting to draft the Introduction.

Soon you will need to add **Keywords** under the Abstract at the beginning of the paper. If you approach the Introduction as suggested in this section, the requisite Keywords covering the subject matter will already be in the front of your mind.

5.3   A Further Example

The broad area, dealt with in one or two sentences at the very start, must lead on to the particular subject of research interest. To give another simple example: if the main area is 'legumes', your particular topic might concern the flowers of one familiar species (point (2) in the box above). Again, a one or two sentence outline will normally suffice. The best policy is to move on quickly to point (3) because you need to identify the problem others might have approached, but not answered, and on which you have therefore decided to put forward your ideas and this has now been explored more thoroughly. This then leaves you to expand in the later sections of the paper on how you went about this business and what results you obtained.

5.4   The Basis of all Research: the Hypothesis

(See the appendix to this chapter for more on the hypothesis.)

Research emanates from a problem needing to be solved. You may have identified a gap in knowledge from your literature survey, or perhaps you can throw light on a poorly understood phenomenon. The problem could have implications for other activities, e.g. related research by groups with which you collaborate, or maybe the search for a new drug for some specific disease in a pharmaceutical company. There can be many reasons, so your job is to explain as quickly as possible why you chose to work on this problem (a small expansion of point (3)). You can do this before you state the problem, but irrespective of order, the problem and the reason for focusing on it must both be stated.

5.5   The Immediate Background

You now need to consider the most recent findings from other researchers, since they will be especially relevant for your reader. If your research has developed out of recent work published by other groups, that work must be outlined. Without dwelling on previously published facts in detail (since this is in the public domain and will already be known by most of your readers), you need to explain how your approach will differ from others and indicate how it could improve our understanding of the problem. Someone may have written a recent review about the issue, or devoted time to it in a longer document. Clearly, a review summarizing and throwing light on recent results is valuable here, but again do not go into detail – it already exists.

In doing the above, you will have begun to formulate your hypothesis ever more clearly and precisely, which should now be presented to the reader as the raison d'être for your research project. A statement of the hypothesis can be incorporated into any part of the Introduction that seems appropriate, but many writers leave it until the last paragraph before the Materials and Methods. You can therefore end with a single sentence (re)**stating the hypothesis** and your methodological approach, clearly and in the most succinct way (see Appendix 5.1). For example: 'We therefore believe that x is responsible for y (**the hypothesis**), and have used major differences in metabolomic profiles to investigate this possibility (**the approach**).'

Remember that the Discussion section will require putting your new findings into the appropriate context (dealt with in Chapter 4). This will entail some duplication of the Introduction, but this duplication must be kept to a minimum, for example by having the general context mostly in the Introduction and the specific context relating to each new finding in the Discussion. Since you have already drafted your Discussion, rereading the Introduction in the light of the arguments presented in it will now allow you to redraft the whole of the Discussion, thereby streamlining it.

Appendix 5.1 The Hypothesis

The Essential Nature of a Hypothesis

The hypothesis is the springboard of your paper. It encapsulates the motivation for the research, and it also tells the readers – both in the

Abstract and the end of the Introduction – that you set out to solve a problem. The hypothesis is the provisional answer to that problem. When writing a thesis, good graduate students will state their hypotheses at the outset, and their mentors must ensure this is done. You need to do the same in a paper. The Introduction should have established enough background for you to articulate the problem that you (or others) have identified and wish to solve. This will enable you to state your hypothesis, ideally in one short (usually final) sentence with minimal dressing-up. The busy reader often goes straight to the last sentence of your Introduction.

Papers without Hypotheses

An alarming number of papers are submitted to journals without a clear hypothesis. Several of the papers I receive are returned to their authors after triage for this reason. Of course, some submissions – fact-finding, data collection and purely methodological articles, for example – have no strict need for a hypothesis, but one can usually be formulated nevertheless. For instance, suppose you wished to identify a suitable substrate for an enzyme and tried 40 different chemicals. This can be seen as rote testing, yet a hypothesis might still be articulated: if you have any information about the active site of the enzyme, you might hypothesize, for some reason known perhaps only to yourself, as to which of the 40 possible substrates is likely to be most suitable (an informed hunch based perhaps on the affinity of two particular chemical groups).

Testing the Hypothesis

Your article should tell the readers what you did in practical terms (or perhaps theoretically) to test your hypothesis. We called this the approach, to be set out in a few short phrases, the main one being emphasized (e.g. using, say, mass spectroscopy). Most researchers set out to 'prove' their cherished idea rather than test it to destruction (if it has a claim to validity it will be sustained after critical experiments – the Popperian principle). In Chapter 3, it was mentioned that data *pro* the hypothesis can easily be selected for communication, whereas neutral or contradictory findings are often skipped or ignored. There is no reason to discount this information since it can be discussed and possible explanations offered. A niggling little fact can undermine a hypothesis, but this is valuable because, among several possible outcomes, the main ones are: (i) the hypothesis becomes refined and can then be re-tested; (ii) it is refuted and you have to formulate a new one. This is how science progresses, so be open about neutral or contradictory data. Other experts are likely to see flaws in your testing of the hypothesis; you might, for example, have applied inappropriate techniques, or used only one approach when you should have addressed the underlying question from several angles, i.e. your interrogation has been too simple. Others might also have other reasons for some of your data not fitting your cherished hypothesis.

Non-Exclusivity of Many Hypotheses

Quite a lot of research is designed to test a hypothesis in a way that suggests there is a single solution to the underlying problem; this can have misleading consequences. First, 'positive proof' for your hypothesis will leave you feeling satisfied, but what you have found might apply only under the conditions in which you tested it, not under others. Second, problems in biology, medicine, sociology and other disciplines tend to be very complex; there are likely to be several explanations, i.e. multiple (non-exclusive) hypotheses are required rather than a single one. Whenever you write a paper you should keep this in mind.

The Importance of Proper and Full Controls in Experimental Testing

I have already mentioned that too few papers start with a clear hypothesis. I also find an unacceptable number of manuscripts in which the hypothesis being tested seems to be supported by good evidence but the work has been inadequately controlled. In the worst cases, controls are completely absent. I use the word 'controls' here because most (yes, *most*) researchers do not appreciate that most experiments need several controls. An example will clarify this. Let us hypothesize that testosterone production by rat testes after weaning alters the stem cell population in another tissue. To test this hypothesis we can adopt several approaches (as suggested above), but – for simplicity – let us find out whether the stem cells are still altered after orchidectomy. The testes in one group of rats are duly resected. The first control would be to give a similar group of rats a sham operation – being anaesthetized, the scrota opened, but the testes left in place before suturing-up. A second group of controls will be orchidectomized but given testosterone injections to check whether in all probability this hormone is secreted by the testes and is crucial for the change in stem cell populations. A third control is for a third group of rats to be taken through the stress of being anaesthetized but not having their scrota opened. The fourth control group would be rats taken to the operating area, not anaesthetized, but, nevertheless, handled and kept away from their cages like the former groups. A fifth group would be animals that remained in their cages, unstressed and undisturbed. The target tissue in which the stem cell population could be altered now needs to be checked some days later: all the rats will need to be killed in the same way and parallel tissue samples taken for analysis. All six data sets will have to be compared (one experimental and five controls). Perhaps you will find that only the fourth control group differs significantly from the experimental group: the sham procedures produced the same result as the orchidectomy, i.e. different sources of stress gave the same outcome. This would strongly indicate that testosterone alone cannot explain the stem cell alteration: the hypothesis is not valid. If only the experimental and the fifth group controls had been done, your results would seemingly have 'provided positive support' for the hypothesis. *Failure to perform all the necessary controls would therefore have led to an erroneous conclusion.* When you write up a paper, remember that controls are crucial and several may be required. It is better to arrive at a well-established result than (as in this example) an incorrect correlation that seems to 'prove' your idea.

I make no apology for this long discussion on hypotheses and their testing. *These matters are crucial for all experimental work and in the writing of papers, especially primary research articles.*

6   Materials and Methods

A research article is *the final product* of an investigation. It is a report that tells the world what you have done and found. On this account, it must be presented in the best possible way so that its message is to the point, clear and succinct.

6.1   General Overview

The Materials and Methods can be one of the more tedious sections of a paper to complete. Nevertheless it must be done thoroughly because it will contain the details that enable a reader to:

- know exactly which procedures were used to obtain the results, and
- repeat them as accurately as possible to verify/refute the findings, especially if he or she had previously used similar procedures and obtained different results. (However, no one can exactly repeat another person's experiments, so differences often arise – some trivial, others more serious.)

In most journals, the Materials and Methods section is divided into subsections. Normally, the first subsection describes the materials, including (in biomedical papers) the cell types or animals used, and the chemicals and their sources. Subsequent subsections describe the experimental and control groups, the procedures carried out (operations, assays, counting and more), finishing with the statistical analyses of the data.

A typical Materials and Methods section from almost any published paper will exemplify the important points, so look at one or two in the journal(s) to which you are likely to submit your manuscript. Because you can easily follow the format and information content, I will not go into great detail in this chapter but point out where extra care is needed and errors are common.

6.2   A Typical Materials and Methods Section

In this section part of a typical Materials and Methods section is shown. Highlights and superscript numbers are used to direct you to the Note for the given example.

An example of a Materials and Methods section

Materials and Methods

2.1 Cells

Human hepatoma (X5) and a cervical cancer (HeLa) cell lines[1] obtained from the American Type Culture Collection were cultured in Dulbecco's modified Eagle's medium (DMEM)[2] containing 10% fetal calf serum (FCS: Hyclone, Logan, UT, USA), 200 mM L-glutamine, 1% penicillin/strepto-mycin (10,000 units/mL),[3] 1 × non-essential amino acids, and 1 mM sodium pyruvate (the chemicals also from Hyclone). Stock cultures kept in a 37 °C humidified 5% CO_2 in air[4] incubator were passaged on becoming 80% confluent. Experimental cultures were kept under the same conditions and used when 60–70% confluent.

[1] In biological and medical papers, cells are often used for in vitro experiments. It is important to note the type(s) of cells used and their source. If cultures are prepared in your own laboratory or from fresh tissue by a service facility, details of their preparation are needed.

[2] Many culture procedures are fairly standard and therefore full details will not be needed. However, modifications must be described in full. There is no need to use upper-case letters in phrases such as Fetal Calf Serum, even though it is immediately abbreviated to FCS. A similar case is Phosphate Buffered Saline (PBS), which should be phosphate buffered saline (PBS).

[3] The use of units throughout a paper should be consistent, and preferably the SI system should be adopted. Some substances are not chemically pure or have no exact molecular weight. If they are used, for example, in a cell culture medium, the quantity present is indicated (usually as a final concentration diluted from a stock) as a percentage, e.g. 10% FCS,

or as mg per ml (which can be written mg/ml or mg ml^{-1}; check the journal's instructions to see whether a particular format is preferred). When a pure substance with a known molecular weight is used, it is preferable to give its concentration in molar units, for example 1.5 M, 1.5 mM, 1.5 μM. These can also be expressed as 1.5 M, 1.5×10^{-3} M and 1.5×10^{-6} M. Be as consistent as possible: try not to use different ways of expressing the same units. If you can use molar concentrations throughout, stick with them. There are other ways of expressing amounts of a substance; one in common use is 'Units', as in '10,000 units/mL' in this illustration. The Units (upper-case U) give the strength of the substance (in this case penicillin), indicating how active it is.

There is an inconsistency in the example: the use of '0.3 M NaCl' is noted, as appropriate, but '0.05% sodium azide' is mentioned later. Sodium azide (Na_3N) is a pure chemical and therefore its molar concentration should have been stated. The use of % should be restricted to substances that are not pure, as in 10% FCS.

Chemicals come from different suppliers; the sources of the less common ones should be declared (see Note 5). Purity might need to be mentioned. If there is evidence that some test substance varies significantly from source to source, and different laboratories have found dissimilar results with this substance, the lot or batch number used should be stated.

There is a further point about the presentation of units in a paper: not only is consistency important, so is the way in which they are annotated. For example, five millilitres can be expressed in the following ways, 5ml, 5 ml, 5mL, 5 mL, and 5×10^{-3} L. Note the space between '5' and 'ml' in the second example. More often 'ml' is used rather than 'mL'. There is no hard rule in most journals, but each has a preferred format for units.

Units can be overused in many papers; here are two common examples:

- 'Samples were taken at 0 min, 1 min, 2 min, 5 min, 10 min, 30 min and 60 min', which is better expressed as 'Samples were taken at 0, 1, 2, 5, 10, 30 and 60 min.'

- 'The incidences were 5%, 7%, 13%, 72% and 98%', which should be given as 'The incidences were 5, 7, 13, 72 and 98%.'

If the number is between +1 and −1, then the decimal marker is always preceded by a zero: −0.234, not −.234.

^{4} Many people write '5% CO_2' without mentioning that it is the concentration of this gas in air.

2.2 Animals

New Zealand rabbits (~2.5 kg) and BALB/c mice (~25 g) were housed in an animal house under strict hygienic conditions at 24±1 °C in which a 12 h light/dark cycle was maintained. Experimental work was approved by the Ethics and Animal Care Committee of Jenkins University.

2.3 Northern blot analysis

The coding sequence of NgRB mRNA was PCR-amplified, gel-purified and ligated to a pEGFP-C2 vector (Clontech Laboratories Inc., Palo Alto, CA,

USA).[5a] The sequence of the insert DNA was confirmed before use as the hybridization probe in Northern blotting. To do so, the insert was released by restriction enzyme digestion (*Bam*HI and *Eco*RI) and the 329 bp NgRB fragment was gel-purified. Rediprime II random primer labeling system (Amersham Pharmacia Biotech, Buckinghamshire, UK, USA) was used to label the NgRB probe and β-actin cDNA internal control probe with [α-^{32}P]dCTP. The probes were purified with a QIAquick PCR Purification kit (Qiagen, Hilden, Germany). Multiple Tissue Northern (MTN) Blots (Clontech) were used to check the presence of NgRB. For the analysis, MTN membranes were soaked for 1 h at 68 °C in ExpressHyb Hybridization Solution (Clontech).[5b] The [α-^{32}P] dCTP labeled probe was denatured at 100 °C for 5 min, immediately chilled at 4 °C for 5 min, and added to MTN-containing fresh ExpressHybm hybridization solution. Hybridization was done at 68 °C for 16 h, followed by 3 washes of 15 min at room temperature with 2 × SSC solution (0.3 M NaCl, 30 mM sodium citrate, pH 7.0)/0.05% SDS, and 1 wash in 0.1 × SSC/0.1% SDS at 50 °C for 40 min.[6] The blot was covered immediately with an Imaging Plate (Fujifilm) for 4 h before the radioactivity was assessed by phosphorimaging instrumentation (Fujifilm FLA-3000, Tokyo, Japan).

2.4 Production of anti-DC2 polyclonal antibody

2.4.1 Covalent conjugation of the polypeptide with its carrier protein[7]

Consensus peptide sequences mouse and human RLYGLMNTSG[8] were designed from their NgRB protein sequence. They were conjugated with mcKLH (murine culture keyhole limpet hemocyanin; Pierce, Rockford, IL, USA) carrier proteins using a ZTM Sulfhydryl Reactive Antibody Production and Purification Kit (Pierce). The resulting mcKLH-RLYGLMNTSG conjugates were stored at −80 °C until used to immunize rabbits.

[5a,b] As mentioned in Note 3, chemicals and other materials used in experiments come from particular companies. It is not necessary to name the very commonly used ones (salt, water, etc.) unless they are special types. The suppliers of the more important ones should be stated by company, city and country, e.g. (Clontech Laboratories, Inc., Palo Alto, CA, USA). After the first mention it can be referred to as (Clontech).

[6] Longer sentences can help when you need to list the steps in a procedure, but they should be used judiciously. One procedure usually follows another in sequence, so there is no need to repeat the words 'then' and 'next'. Take 'We loaded 5 tubes with sample solutions, then spun

them at 300 g for 5 min, then removed the supernatants, and then stored them at −20 °C.' This is better written 'We loaded 5 tubes with sample solutions, spun them at 300 g for 5 min and removed the supernatants, which were stored at −20 °C.'
[7] It is useful to distinguish different parts of a section by using further subsections. Here, the second heading under Materials and Methods is 2.4, which is subdivided into 2.4.1, 2.4.2, and so on. The title of each subdivision may be indented and the font smaller and italicized to distinguish it from the major sections.
[8] Acronyms and abbreviations can present problems. The item in the text here with superscript no. 8 comprises single-letter symbols for amino acids in a string (a peptide). Accepted symbols do not usually have to be explained, and some acronyms are of this type (e.g. RNA and DNA), but others need to be given in full before an acronym is assigned (e.g. monosodium glutamate, MSG). This latter example shows how such an acronym should be introduced. It is unwise to revert to the full name thereafter; stick to the acronym. It is also wise to check whether the journal to which you plan to submit prefers the less common abbreviations and acronyms to be appended in a list somewhere in the paper (e.g. just below the abstract). SDS-PAGE (sodium dodecyl sulphate polyacrylamide gel electrophoresis) is another case in point, seen lower down in this example.

2.4.2 Rabbit immunization

Before immunization, pre-immune sera were collected from New Zealand rabbits. Purified mcKLH-RMKLPGYLMG conjugates (500 µg) were first dissolved in PBS before mixing to form a stable emulsion with an equal volume of complete Freund's adjuvant (Sigma). Rabbits were injected subcutaneously into 4–5 sites on the back and sera were collected 2 days after immunization. After 2 weeks, the rabbits were immunized with fresh batches of conjugate:adjuvant, and twice more (250 µg) at 2-week intervals. Sera were collected within 2 days after the last treatment.

2.4.3 Antibody purification

Sera from rabbits were mixed with sodium azide (0.05%) and centrifuged at 15,000 g at 4 °C for 5 min. The supernatant was diluted with an equal volume of TBS followed by filtration through 0.2 µm Acrodisc syringe filters (Pall Life Sciences, East Hills, NY, USA). The serum mixture was passed through a Protein G Sepharose-4 Fast Flow column twice at 2 mL/min (Amersham, Uppsala, Sweden). Antibody was eluted with 15 mL of elution buffer (50 mM glycine-HCl, pH 2.7). Aliquots (1 mL)[9] of the eluate were collected into microcentrifuge tubes containing 100 µL of neutralization buffer (1 M Tris-HCl, 1.5 M NaCl, 1 mM EDTA, 0.5% sodium azide, pH 8.0). Aliquots with high protein absorbance (A280) were pooled, pH adjusted to 7.4, and stored at 4 °C.

[9] It is considered undesirable to start a sentence with a number, although it is not a sin. To obviate the need to do so, the device illustrated in the text can be used: start with a noun and put the relevant units in parentheses after it. Other examples would be: 'New medium (20 ml) was added after...' or 'Twenty millilitres (of *x*) was added...'

2.4.4 Analysis of antibody specificity by RNA interference

HeLa cells were seeded on 6-well culture dishes (2.5×10^5 cells/well) and grown overnight. Initially, $2\,\mu L$ DharmaFECT Transfection Reagent (Dharmacon, Lafayette) was mixed with $98\,\mu L$ serum-free/antibiotic-free DMEM and incubated at room temperature for 5 min. To this was added either $50\,\mu L$ 1 × siRNA buffer (Dharmacon) (Control group I, CI), $50\,\mu L$ of $2\,\mu M$ siCONTROL Non-Targeting siRNA (Dharmacon) (Control group II, CII), or $50\,\mu L$ of $2\,\mu M$ NgRB pooled siRNA (Experimental group).[10] The mixtures were incubated at room temperature for 25 min. HeLa cells were washed twice with serum-free/antibiotic-free DMEM before transfection. This supplemented DMEM (2 mL), containing either control or experimental DharmaFECT mixtures, was added to cultures and incubated for 7 h before 2 mL of[11] DMEM containing 20% serum without antibiotics was added. Cells of the experimental group were collected at 24 and 48 h, and control cells were collected at 48 h post-transfection. Antibody specificity was examined[12] by Western blotting.

2.4.5 Analysis of antibody specificity by in vitro transcription and translation assay

DC2 protein was produced in a cell-free system using the TNT T7 Quick Coupled Transcription/Translation System (Promega, Madison, WI, USA) following the manufacturer's protocol.[13] In brief, the plasmid containing the coding sequence (CDS) of NgRB pCDNA3.1/NgRB2 (verified by DNA sequencing) was linearized by restriction digestion with *Eco*RV. The linearized pCDNA3.1/NgRB2 containing the CDS of the NgRB gene was gel-purified and dissolved in $20\,\mu L$ nuclease-free buffer, and mixed with $40\,\mu L$ of TNT Quick Master Mix, $1\,\mu L$ PCR enhancer, and $1\,\mu L$ 1 mM methionine (Promega). After 2 h incubation at 30°C, the in vitro TNT product of the NgRB protein was used to verify the specificity of the anti-NgRB polyclonal antibody.

[10] To identify the different groups, populations etc. that are going to be compared in a paper (usually the experimental subjects and their controls), they can be listed in the running text. However, a neater way is to assign numbers or letters to the groups and list them in an indented paragraph. In a good experiment with many experimental groups, there will also be many control groups (see Chapter 5, Appendix 5.1) – some treated with placebos instead of the agent (say, a drug), others given neither agent nor placebo (untreated). It is best to check how many control groups you need. They all need to be reported in the paper.

The amount of detail you should provide about the sets or groups in each experiment depends on the nature of the work, and few generalizations can be made. Some people put a lot of details into the Materials and Methods section, whereas others give them in the Results as each experiment is described, or in the figures and tables. At minimum, the Materials and Methods must clearly distinguish the experimental groups.

[11] This note concerns the use of the word 'of'. It is commonly used in this situation but should be omitted: e.g. 'we added 5 ml of MGX before...', where it should be 'we added 5 ml MGX before...'.

[12] In many papers, one particular word (verb) is repeated in describing what was done and what was found. For example, we treated group X with Y ... we treated a second group P with Y and D ... we treated group b with only D. Or we evaluated the instrumentation before use in the experiment ... the levels of haemoglobin were evaluated every second hour... and the results were evaluated after normalization. Repetition of a word with limited meaning is jarring and unnecessary. Science is all about measurement. In English, many words can be used to describe how data were obtained or results interpreted. We can measure, assess, evaluate, calibrate, enumerate, and so on; but each of these words has a very precise meaning in its particular context in English. A couple of popular words today are 'performed' (procedures of all types, even statistical analyses, are performed – in English to perform is to act, usually to an audience!). 'To perform an analysis' is simply 'to analyse'. ('Western blot analysis was performed according to Chen et al. ...' would become 'Western blots were analysed by the method of Chen...'). The second word is 'demonstrate' – we demonstrated..., when showed, found, etc. would be better. To add a little drama to scientific papers, emotional words are introduced: e.g. We were 'surprised' to find.... Perhaps the most common in papers on animal experiments is that the mice were 'sacrificed', when the word 'killed' is accurate. Sophists might argue that the animals were indeed sacrificed on the altar of science, but the animals were nevertheless killed. Even 'euthanized' is an equally unattractive word.

[13] This reference to carrying out a procedure 'according to the manufacturer's instructions' has become commonplace in papers. If you are using a kit or some special apparatus, it would be folly to use it without following the maker's instructions. Therefore it becomes unnecessary to use this phrase. It would be important to mention, however, that some deviation or modification of the manufacturer's protocol was introduced.

2.5 Western blot analysis

Western blot analysis was performed according to Chen et al. (2007).[14] Cells were collected in phosphate buffered saline (PBS), centrifuged at 2000 g, and the pellets lysed[15] in Tissue Protein Extraction Reagent T-PER (Pierce). Animal tissues were collected in T-PER and ground with a tissue homogenizer. Cell-free in vitro[16] translated NgRB protein was used without further treatment. The samples were incubated on ice for 40 min followed by centrifugation at 15,878 g[17] for 20 min, and the supernatant was stored at −80 °C. Protein concentration was determined with a BCA Protein Assay Kit (Pierce). Protein samples (50 μg) were subjected to 12% SDS-PAGE under reducing conditions on 5% stacking and 12% resolving gels, before being electrotransferred to a nitrocellulose membrane. NgRB protein was recognized with its rabbit antibody (1:200 dilution) and either a mouse anti-actin antibody (1:5,000 dilution; Chemicon, Temecula, CA) or a mouse anti-tubulin antibody (1:5,000 dilution). Western blots were developed with the Enhance Chemiluminescence Detection System (Amersham). Chemiluminescent signals were captured[18] using the Fujifilm LAS-3000 system (Fujifilm, Tokyo, Japan).

---

[14] Increasingly, references to other papers are used to circumvent the need to provide full details of an experimental procedure. This saves space, but the reader who wants to know more about the procedure has to work harder. Sometimes the example above – 'by the method of (ref.)' – can be reduced to something like 'The MMT method was used to measure viability (Jones, 2000).' It is appropriate to refer to methodology papers in this way, but it is not good practice to cite experimental work published by others as justification for a method you have used, unless this is absolutely necessary.

[15] Listing a procedure succinctly within a sentence enables easier reading, and one can infer much when what is written relates to common laboratory practice. Preparing cell pellets is an example; one can easily infer that the cells had to be spun in a centrifuge! What should be avoided is the use of 'then' and 'next' as each procedure is mentioned in a sentence or listing; the reader can easily infer them from the text.

[16] Latin words are sometimes put into italics, but this is not essential. Indeed, it is now common not to use italics for 'per se', 'in vitro', 'inter alia', 'et al.', just as we already do for 'i.e.', 'e.g.' and 'etc.' (ie can be used just like i.e., but stick to one of the two alternatives). Use of Latin, as in these examples, tends to be in abeyance – making it odd that 'via' has become so popular when 'by' is the simpler English word.

[17] Extraordinary degrees of accuracy are often assigned to figures in scientific papers. It is highly unlikely that the authors in the example measured the rpm of the centrifuge at 15,878; 16,000 would be accurate enough. You often see values such as 51.072 given for a measurement where the tolerance of the instrumentation is 1–2 per cent, making 51 good enough and 51.1 more than accurate. There are cases where extraordinary accuracy can be reported, as in astronomical measurements made with exceptional instruments. However, in biomedical sciences, although instrument accuracy might be high, the material being measured is inherently variable, so the data have limited accuracy.

[18] We are back to odd usage of English words. 'Captured' (like 'sacrificed' in Note 12) is now frequently used for recording images, but in plain English, people 'take' pictures. Capture has a connotation of taking hold of something that is on the loose.

2.6 Heat shock treatment of cell lines and animals

HeLa cells were cultured in 10 cm dishes sealed with 2 layers of parafilm and placed in a pre-heated (37 °C) water bath for 20 min, after which half the cells were frozen at −20 °C. The other half was continuously cultured for 30 min before being frozen for follow-up experiments.

Mice were anesthetized by ip[19] injection of 25 per cent urethane (30 μL/g) and immersed so that only the head was above the water surface. They were left for 5 min after their anal temperature had reached 41 °C. The mice were killed by cervical dislocation, and the organs were collected at 0 min, 5 min, 10 min, 30 min, 60 min, 120 min, 180 min, and 240 min after heat shock treatment.

2.7 Statistical analysis

To compare relative protein expression in the different groups, one way ANOVA was performed and means were compared using Tukey's Multiple Range Test using SPSS. Significance was set at $p < 0.05$.

[19] The question here is whether shorthand for intraperitoneal (ip or i.p.) or subcutaneous (sc or s.c.) is permissible. They are familiar abbreviations in biomedicine that are rarely misunderstood, so it is quite acceptable to use the shorthand versions.

6.3   Concluding Remarks

Most of what needs to be said about Materials and Methods has been covered in these annotations. In another chapter we will discuss the importance of stating that *ethical approval* had been obtained from the appropriate authorities for procedures involving living animals. Most countries require investigators to be trained and gain certification for experimenting on live animals. In medical research involving human beings, not only must approval be obtained from the medical authorities, but written consent from the subject is almost mandatory. The name of a subject or a patient cannot be mentioned or any details given that might enable that person to be identified.

If any materials, gifted or not, were obtained from another person, he or she should be mentioned. This can be done in Materials and Methods or the Acknowledgements at the end of the paper. Services of facilities used in the work, e.g. blood sample analyses in a central laboratory, ought to be mentioned, particularly if they were freely given. Many databases are now available for protein and DNA sequences, enzymes (nomenclature and EC number) and so on. Where use has been made of such a service, usually free on the Internet, you should state which it is and give its reference number (e.g. Blast; X1234).

Any hazardous chemicals or operations should be noted with regard to the safety precautions, so that a reader using the same materials or methods is forewarned of danger, and can secure appropriate health and safety permission (e.g. where a contagious virus is being investigated).

A considerable amount of discretion is needed in deciding what to include and what not to include in this section. If in doubt, do not leave it out! You can also seek advice from editors, who might ask for more details, for example following referees' reports that have drawn attention to gaps in the information provided in a manuscript. And do not forget to run a final check. In the Results section, you might have included (often on second thought) more experimental data than you originally intended, making it necessary to add more information about the materials and/or methods. It is therefore wise to go back over the paper on completion of the draft and ensure you have included all the necessary details.

The positioning of the Materials and Methods section differs among journals; it is usually either after the Introduction or after the main body of the text. In some cases a smaller font is used. There is also the possibility of using Supplementary material (including online files to which links have been made) to expand on complex methods, adequate description of which would otherwise require a lot of space in the main text.

Figures and *images* are not usually included in the Materials and Methods section unless they are necessary to expound a particular point (e.g. a flow chart or schema of an elaborate procedure).

Finally, every effort should be made to keep the Materials and Methods as short as possible – although, to ensure all the important points are made, it should not be pruned too vigorously.

7   The Abstract
The Summary of the Main Findings

A research article is *the final product* of an investigation. It is a report that tells the world what you have done and found. On this account, it must be presented in the best possible way so that its message is to the point, clear and succinct.

7.1   Purpose of an Abstract

Creating a good Abstract is one of the most challenging tasks in writing a scientific paper. Several important points must be made in a very limited space. Having now read six chapters about paper writing, you will have all the key features of your manuscript in mind. In the Abstract you must present those features clearly and succinctly to your readers. It is imperative to *keep the reader in mind*. You do not write a paper to satisfy yourself about the work you have completed, but to apprise your readers/audience of that work.

7.2   Structured versus Non-Structured Abstracts

When you have decided to which journal you will submit your manuscript, check whether the instructions to authors demand a structured or an unstructured Abstract. Most journals have unstructured Abstracts, but others specify subsections (typically Background, Methods, Results and Conclusion). The relative importance placed on subsections in structured Abstracts can vary, which gives unstructured Abstracts an advantage (see below). Structured Abstracts are usually allowed to be longer than unstructured ones, but check the length limit in the Instructions to Authors. In any case, it is best not to exceed 150–200 words. This is why writing an Abstract can be so difficult – but it is of prime importance because many readers will see little (if any) more of your paper. Readers may want to know the gist (essence) of your work, but will have little need, or time, to read the whole article. In the Abstract, therefore, you should go straight to the main point.

Abstracts

Abstracts need to convey the main message with the least introduction possible. Word limits on Abstracts are very restrictive; there is room for nothing except *the most essential information.*

7.3   Importance of the First Sentence

Although many writers would wish to start their Abstract by setting out the background, as required in a structured Abstract, this can detract from the important message, viz. the result of testing your hypothesis. If the hypothesis is stated in the first sentence, the background to it will be apparent to the interested reader (who, remember, may be as knowledgeable as you about the subject matter intimated by your title). If your research led to a clear conclusion, it can even be worth starting with that conclusion. For example: 'In examining the relationship between x and y, it is now clear that the former is inversely related to the latter, but only within the pH range 6.1–9.2.' Cognoscenti reading this will know something about the background and previous findings. Such readers will therefore understand the hypothesis being tested and will probably guess which procedures have been used to establish the conclusion, making it less important for you to describe the background and methods.

My preference is to get the most important message, the main result, across at the very start. Like a headline, it will immediately grasp the reader's attention. It can be followed by a brief mention of the hypothesis being tested, then some indication of the motivation for the work (the background can be evoked here); perhaps previous studies failed to approach the problem in the right way and therefore started from a different hypothesis. Other important findings potentially supporting the first statement can then be mentioned – briefly, not in detail. Actual figures from the Results section are not normally included in the Abstract, though general statements can be made. For example, if the value of x was 90 ± 7 but fell under certain circumstances to 29 ± 5, you might say that it fell to about a third of its original value – if this is an important message of your paper.

7.4   Building on the First Part of the Abstract

By now you have either stated your hypothesis and then your principal findings, or vice versa, so you will need to expand a little on the context. The hypothesis could be entirely new, or it could relate to another hypothesis, which might need to be explained. The findings are the real message of the paper. They should resolve a problem or perhaps describe a new phenomenon; in other words, identify something new and original – otherwise they can do no more than corroborate previous work. Presenting the Abstract should bring your main message into sharp focus.

If something new is being reported, as one should expect of a scientific paper, the expression 'for the first time' is redundant. We frequently read sentences such as 'the data we report show *for the first time* that x is in fact inversely proportional to y'. This expression is commonly used by authors who wish to make the prior claim to some finding, but it is unnecessary. Any paper imparting significant new data is important in this respect – by definition. There is no need to state the obvious.

Stating the hypothesis

The hypothesis is the reason for the work and the findings being reported. It is therefore essential to include it, in its most succinct form, in the Abstract.

7.5   Putting in Some Background

It can be difficult to judge how much background is required in an Abstract, but whatever the situation, strive to keep it to a minimum. The subject matter and the results to be reported largely determine the amount of background necessary. If something entirely new is to be presented, little or no background will be available; if the subject matter is already familiar to readers, there is little point in spelling it out. However, when a topic has recently been addressed by many research groups, the immediate background becomes more important. In such cases, the explanation of the phenomenon under study is likely to be controversial; for example, different groups might have approached the problem from different angles and no consensus has emerged. The easiest plan is to state the dilemma as simply and briefly as possible and indicate how you have addressed the problem. Further information can be given in the Introduction and the arguments can be explored in the Discussion.

The awkward problem of background information in an abstract

In a structured Abstract, a background section is expected, but in an unstructured Abstract there is often little need to include background. The hypothesis might in itself indicate the nature or context of the problem.

7.6   Avoidance of Citations

Conventionally, citations (references) are avoided in an Abstract. The Abstract states the most important findings simply; the relevant literature will be cited in the rest of the article. There may be a case for adding a citation where, for example, some highly significant and long-established work has come under scrutiny and new findings that challenge its validity are to be reported. Thus, if the central dogma of molecular biology were to be challenged by new findings, it could be important to cite the paper in which the central dogma was first introduced. But as you will appreciate, such a reference would be unnecessary for most readers, who will know the central dogma and its provenance.

There could be a case for citing a very recent paper reporting an important finding that your work either challenges or corroborates, since interested readers might not be aware of it. However, even in this case there are ways of introducing your findings in the Abstract without the citation. For example, you might say 'We herein propose an alternative hypothesis to the recent proposal that x is inversely proportional to y,' obviating the need to identify the relevant publication.

7.7   Concluding an Abstract

The conclusion of an Abstract has to do two things: first, it must re-emphasize your main finding and, second, it must indicate consistency (or otherwise) with received wisdom and/or the contribution made to the sum of knowledge – in effect, how your work is to be seen in the greater scheme of things. It does not need to suggest what further research is needed. Anyone who understands the rest of your Abstract will realize that this is the present 'state-of-the-art'. Research is a never-ending process; it always leads to more experimentation and further findings. Most people will know where your paper is leading from its relevance to the state of present research in that area, making suggestions about future work unnecessary.

7.8   Keywords and Abbreviations

Just below the Abstract you will need to choose (preferably from the Abstract) about six to seven **Keywords** that will help readers to understand the thrust of the paper. Just below the keywords, a list of **Abbreviations** used throughout the article is often required, especially if they are numerous or not widely known. Accepted abbreviations (those that are regularly used by scientific writers, e.g. DNA, RNA and ATP) need not be included, so only list the less familiar ones. A word that you will subsequently abbreviate is given in full at first mention in the text of the article, with the abbreviation in parentheses after it, e.g. diphenylhydrazine (DPH). You should be consistent, i.e. continue to use the abbreviation in all subsequent mentions; do not alternate it with the full version.

Many accepted abbreviations relate to units and frequently used words, such as h for hour(s), min for minutes, g for grams, and rpm for revolutions per min. These are sometimes listed in the Instructions to Authors of a journal, so you should check this and use the abbreviations recommended by the publishers. Again, be consistent and do not switch between the short and long versions. If there is any doubt or ambiguity, it is best not to abbreviate.

Some journals place these two sections in footnotes rather than inserting them below the Abstract, but you should proceed as described in this chapter. The publisher will determine where to place them when the final version of the paper is made ready for publication.

In the next chapter we will deal with the remaining subsections of a typical paper. To prepare for this, look at a published paper (the one in Chapter 2 will serve for this purpose) and you will see short sections entitled: Conflict of Interest, Acknowledgements and References.

8   The 'Smaller' Sections That Complete a Paper

> A research article is *the final product* of an investigation. It is a report that tells the world what you have done and found. On this account, it must be presented in the best possible way so that its message is to the point, clear and succinct.

8.1   Other 'Small' Sections Required

We have to complete a draft of a paper by adding some small sections. Some of these are occasionally omitted, but if you include them you will facilitate the progress of your paper towards publication. Some are obviously essential (**Title** and **References**), but I also refer here to **Conflict of interest, Authors' contributions** and **Acknowledgements**, which are sadly sometimes overlooked. The small sections are therefore:

- **Title**
- **Authors, affiliations and contributions**
- **Corresponding author**
- **Conflict of interest**
- **Acknowledgements**
- **References (Bibliography)**

8.2   Title

This has been mentioned in previous chapters, but its importance for the completed manuscript is self-evident. We have noted that it is best to get everything else in order, including the abstract, before assigning a Title.

A paper develops as you write it, and it is only at the very end that a suitable title will have formed in your mind; it is the finely honed (refined) result of this sustained thought. It needs to indicate *unequivocally* what the paper is about, without being too long. It can indicate a new approach to the problem in hand without stating your main conclusion, which will be appreciated as soon as the reader looks at the Abstract. Most search engines for scientific papers retrieve the Titles and Abstracts of all articles in their listed journals. Search engines also lock on to the *keywords*, and some of these ought to feature in the title.

Many present-day authors like to state their major finding boldly in the title; this is referred to as a declarative title, but this will not necessarily draw a reader into perusing the article more thoroughly. Indeed, they may take the title as a 'fait accompli' – they have got the message and will not even bother to examine whether the evidence within the paper substantiates the claim. This can be quite dangerous, especially if the paper is weak in the evidence it presents. A newsvendor will not sell many papers if his hoarding gives the full message, e.g. President calls for calm! About what? Buy the paper and read on!

A title does not have to start with the definite or indefinite article ('The' or 'A'). In previous times, papers were catalogued using the first word of the title, so 'The Effect of X on Y' would be classified under T. Thus it is better to start with 'Effect of X on Y' – it also saves a word! If 'Effect of X on Y' has been the subject of previous publications, your title needs to be more specific, e.g. 'Effect of X on Y under reduced hydrostatic pressure.' Open the Abstract with a sentence that does not repeat the title; repetition always annoys referees, editors and readers. These days the word *effect* seems to have given way to *impact* in most texts of scientific papers, but oddly enough it is seldom used in titles!

Titles

The title is the first thing a reader will see. You want those most interested in your subject matter to read on, so it has to be attractive and appealing. In common parlance, it needs to have 'punch', more like a newspaper headline!

8.3   Authors, Affiliations and Contributions

The **author list** comes just beneath the title and generally has a specific order, although this varies. An old but now seldom-used practice is to list the authors' names in alphabetical order of surnames. Often the senior author, i.e. the person who is mainly responsible for the research, comes first, and those who contributed less are listed in descending order of input. It is equally common practice for the principal investigator (PI) in whose laboratory the work was done to put his or her name last, particularly when they have become well-established in the field of study.

Rivalry can develop in author preference; it is best to sort this out amicably at an early stage, even before you start drafting the manuscript. If someone has initiated and done much of the research, e.g. a post-doctoral worker, the PI may decide when there is sufficient new information and understanding to begin writing. Between them, they should agree on the co-authors to be included and their order. In some cultures there are different priorities; in some countries, the chief of a laboratory or an institution often insists on being the first-named author or at the very least being included, even if his or her contribution is only in administering the laboratory and department in which the work was done. There are no set rules, but authors can be upset if they are placed lower down the list than they believe their contribution warrants. There have even been cases where the most import-ant contributor has left a laboratory, and later finds a paper about the work has been published without his or her involvement. There is no easy solution to this problem, but it needs to be very carefully considered by all concerned. The problem is compounded when the work is reported from several different laboratories and institutions, but this is not a complication into which we can delve here. The opposite is totally unethical, i.e. a person leaves a laboratory where a collaborative project involving several workers has been completed, and then proceeds to publish it from elsewhere as if it were his or her own work. This can lead to unpleasant repercussions and sanc-tions affecting the future prospects of the perpetrator.

Always check either a copy of the journal or the Instructions to Authors to find how to list each author in the affiliations. In some journals 'F. Strong and D. Brave' might be used, while others will require 'Strong F, Brave D'. When full names are used (e.g. Frederick Strong, David Brave), it might not be clear to all readers which is the forename and which is the surname (family name). The conventional order differs from country to country (compare UK names with those from Hungary or China).

A problem with author names

Countries across the world differ in the way names are given. Some, e.g. Hungary and China, use the reverse order from Western cultures: e.g. 'Sarkadi Attila' instead of 'Attila Sakardi' (which would be the Western convention). It is essential to identify the family name or surname by underlining or bold type in order to eliminate ambiguities that could confuse the publisher.

The best way to give the **affiliations** is to use superscript letters or numbers for each author, e.g. F. Strong[a], D. Brave[b] (alternatively [1]Strong F, [2]Brave D). A list of the superscripts follows immediately below the author list, any two or more authors having the same affiliation sharing the same superscript. If an author has moved from the affiliation to a new address during publication of the paper, a further superscript can be added using a symbol (e.g. if letters have been used so far, then [+] might be added to the superscript of the author who has changed affiliation – [b+]Brave D), and a note stating this new address is added below the affiliation list. Affiliations should comprise the laboratory (department), institution, city and country. However, full address details are not essential for every author. They are needed for the corresponding author (see Section 8.4), but it is customary to give the email addresses of all the authors.

Authors' contributions to the paper are commonly stated in papers nowadays, especially when two authors have made similar major contributions. Not all journals require this information, but there is no reason not to include it. Exactly where it will appear in the published paper is not your concern; it might be on the title page or after the Acknowledgements, which will be dealt with by the editors and publishers of a journal.

8.4   Corresponding Author

It is preferable to use only one person as the corresponding author, denoted by an extra superscript symbol in the author list; in this case it might be *. The same superscript is usually put at the bottom of the title page, followed by the corresponding author's name, full address and contact details (telephone and email). Exactly where the corresponding author's details appear in the final printed publication is again a matter for the publisher to decide.

The corresponding author – often, but not always, either the senior author or the PI – is the person through whom all communications with the editors and publishers take place. Thus he or she is responsible for all the other authors and needs to know every detail about them and the paper. If there is any contention concerning authorship or content of a paper, editors can also insist that all the authors are contacted, which is a good reason for including the email addresses of all the authors at submission.

8.5   Conflicts of Interest

A **conflict of interest** arises where an author has a vested interest in the subject matter being reported. A typical case is where one or more of the authors have an involvement other than the purely scientific one related to the publication. If a new drug has been the subject of experimentation, authors might hold a stake in the biotechnology or pharmaceutical company concerned or are being paid as advisors or assistants to help in drug development. There can be other sources of conflict, too diverse to be discussed here, usually stemming around *intellectual property* (IP). However, if there are any doubts, especially where the ownership of the IP is concerned, it is better to resolve them before submission. This might require advice from specialists (e.g. advisors on IP, notably legal advisors where IP might lead to the application for a patent). Once again, it is better to be safe than sorry. Big problems can occur after some of the information (e.g. part of the IP) is published if proper care is not taken, such as the loss of the opportunity to patent a drug because details have already entered into the public domain.

A note of caution on the 'ownership' of research data

Once information has entered the public domain by publication in a journal, the rights of any person or company trying to protect it for commercial gain are compromised. Conflicts of interest are always better expressed than omitted from any document proceeding towards publication. If you do not want the data to become public property that anyone can use, you should *not* publish them at this time, but seek legal advice on ways of protecting them (e.g. having the new findings patented first).

The usual place to find statements regarding conflicts of interest is towards the end of the paper, sometimes just after the Acknowledgements. Again, the exact location will be decided by the publisher; it might even be on the title page. (Many other legal and ethical aspects of scientific publishing will be discussed in later chapters.)

8.6   Acknowledgements

All too often the **Acknowledgements** section is omitted by authors; journal editors and publishers should insist upon it, usually as a section immediately following the Discussion. Its importance is at least two-fold: thanking contributors to the work who have not been included as authors, and recording the funding source(s). Omitting the first is discourteous, and omitting the latter is irresponsible and unethical. The following details generally need to be included:

- Names of people who have contributed significantly to the work, such as technicians, secretaries and people who have given advice on your paper; these can include statisticians, supervisors, experts from outside your institution and colleagues who have read and commented on your draft paper. It would be difficult to provide a comprehensive list.
- Gifts from other people or laboratories that have been integral to the work. Similar acknowledgements are sometimes placed in the Materials and Methods section; it is unnecessary to use both.
- Any other outside organization that has done work and provided data on a gratis basis (i.e. not paid for by the grant). For example, you might have had blood samples analyzed at some central facility in your own or another institution.

- Names of the grant-giving bodies, with specific grant numbers or details of the funding source if not an organized governmental body or recognized charity. Where the work has been supported by private donations, a public company or similar body, details are required (check here for any possible conflict of interest). **Funding** is now more commonly being put as a separate section within the Acknowledgements.

If in doubt, do not leave someone or something out. It is better to include as many as you think deserve recognition than to discover later that someone is displeased at having been overlooked. The following box gives a typical example of the three sections appended after the Discussion section of a paper – Conflict of Interest, Authors' Contributions and Acknowledgements.

Competing interests
The authors declare that they have no competing interests.

Authors' contributions
HT and TS performed the research, analyzed the data, and drafted the manuscript. YF helped with cell culture, transfection, immunostaining and Western blotting techniques. NI prepared the retinoid used in these studies. MW helped with immunostaining techniques. MO, HM and SK designed the research, interpreted the data, and revised the manuscript. All authors approved the final version of the manuscript.

Acknowledgments
We wish to thank Dr Strong for his helpful discussion on this paper, Dr Brave for providing us with antibodies, and Mr Yin for carrying out the analysis of blood samples.

Funding
This work was supported partly by Grant-in-Aids from the Ministry of Education and Science (20790215, SK; 18905016, HM; 36790266, HT), a grant for the 'Chemical Genomics Research Program' (to SK), and a 'Special Grant for Promotion of Research' (to HT) from OBSCB.

8.7   References (Bibliography, Literature)

8.7.1   General Introduction

The **Reference** list is an essential section at the end of the paper. The rules concerning its presentation differ greatly among the world's science journals. A huge number of issues and details could be given, but I will discuss those that are probably most important.

Examples of reference lists have appeared in earlier chapters. You will have noted that there are two main citation methods and reference-compiling systems. The first is the **Harvard** (or author–date) system, in which the authors and dates are inserted in the text wherever a publication is cited. These citations are compiled as a reference list in alphabetical order. The second is the seemingly simpler numbered or **Vancouver** system, in which the citations in the text are numbered sequentially. Numbers should appear in text as either non-superscript numbers in square brackets before punctuation, or as superscript numbers following punctuation (reference [3], or reference,[3]). If the same reference is cited more than once, it is given the same number. In the bibliography, the references are compiled in numerical order, but the system used by a particular journal has to be adopted; it cannot be the author's choice. Annoyingly, papers arriving on the editor's desk often do not comply with the system used by the journal. Either the author has not read the Instruction to Authors (perhaps not even bothered to see a paper from the same journal) or the paper has been submitted elsewhere and the authors have not revised it before submitting it to another journal. Never should the two systems be used within the same paper.

8.7.2   Advantages and Disadvantages of the Two Main Citation Systems

The advantages and disadvantages of the two systems are listed, but they are often seen quite differently from the author's, editor's, publisher's and reader's points of view.

The advantages of **Harvard** are:

- An interested reader will probably know the reference from the name of the authors of a citation and the year of publication, so they will not have to check it in the reference list as they read the text. They will therefore follow the writers' train of thought better.

• It is much easier to find a particular reference in an alphabetical list than a numerical one.

Some disadvantages are:

• It takes longer to insert the reference.
• It may be more complicated to add extra information when several citations are given at the same point in the text (as will become clear below).
• It takes up more space than inserting a number.

The advantages of **Vancouver** are:

• It is much simpler to insert a sequence of numbers, saving text space (and so reducing the word count).

Its distinct disadvantages are:

• Even knowledgeable readers can only guess at whose paper is being cited, so if they need to know who exactly is being cited at some point, they have to keep going to the reference list at the end of the article.
• The author has to be careful when the same article is cited more than once, otherwise multiple numbers are attached to the same publication and the reference list is lengthened by redundant repetitions.
• This system is not kind to authors if at any time, notably during revision of a paper prior to full acceptance after peer reviewing, other references have to be inserted.
• The appearance and positioning of the numbers can vary; e.g. (1), [1] and [1], which means that you should adopt the type required by the publisher.

8.7.3   Requirements of Text Citations Using the Harvard System

Harvard style references require more work as the authors and years cited must be inserted. If there is only one author of a cited paper the task is simple, e.g. (Strong, 2007). When there are two authors it would read (Strong and Brave, 2007; or Strong & Brave, 2007). If there is more than one citation to Strong and Brave in the same year, a letter has to be placed after each, as in (Strong and Brave, 2007a, 2007b, 2007c). If multiple references are used at a single point they can be in chronological order, e.g. (Strong and Brave, 2007; Post and Pillar, 2008; Jenkins, 2009), alphabetical order (Jenkins, 2009; Post and Pillar, 2008; Strong and Brave, 2007) or (more rarely) in order of importance. A mixture of the three should never be used. When there are multiple authors in a single citation, the abbreviation et al. (for *et alia*, Latin for 'and others') is used, as in Strong et al. (2008).

8.7.4   Producing a Reference List (Bibliography/Literature)

Modern software (which the author has already used to form a database of references required for the manuscript) inserts a coded reference for each one placed in the text. These can be recovered as a complete list when writing the text is more or less complete and can be sorted usually automatically into the order required by the chosen journal (as discussed above for the two different systems). The software is versatile: if the author checks the format required by the journal to which the paper is going to be submitted, each reference can be converted into the correct presentation provided it was first recorded in full detail. References in the list normally include all the authors (although some journals allow one to add 'et al.' after a certain number of authors). This is followed by the title of the cited paper in full. However, some journals, such as those with restricted space, do omit titles. The name of the journal follows, in full or in an accepted abbreviation (e.g. *Journal of Theoretical Biology* or *J. Theor. Biol.*). The way in which the volume, issue, pages (preferably first and last), and year are included will also have been prescribed by the journal. The order of the individual items differs a lot; e.g. the year of publication might be placed immediately after the authors or at the end of the reference. The software can also provide the correct typefaces and font sizes for different parts of the reference, e.g. the title of an article or book might be required in italics. It is traditional in most journals, for example, to have the volume number in **bold** font, e.g. **12**. Some journals also prefer to include the issue number, e.g. **12** (6) to denote the sixth issue of volume 12. Page numbers for each reference can be those of the first and last pages, or just the former, e.g. **12** (6) 27–35 or **12** (6) 27. Perhaps the most awkward thing to do when compiling references without the aid of the most suitable software is not only to get the right fonts, but the right spacing and punctuation.

There are numerous rules for the layout of reference lists, but the essential points are covered in the foregoing. In some frequently produced journals with many papers and limited space, such as *Nature*, references may be pruned to only the first author followed by 'et al.' (if there are two or more authors); the title may be omitted, and only the first page number and the year given. Although publishers will do their best to help authors produce the definitive version they require for publication, it remains the responsibility of the authors to comply as far as possible. In some journals published online, authors may find their 'final accepted version' returned innumerable times to make them comply with all the requirements.

A quick note on the Reference list

Keep it as short as possible. Check that all your key references are included, with a quick look at online databases to ensure a new one has not just appeared. Avoid citing many references individually if the same end can be achieved by directing the reader to a recent good-quality review.

In previous chapters, it was mentioned that in a primary research article it is best to cite only the essential references, not a comprehensive list of previous publications (you are not writing a critical review paper!). Recent topical papers are important. However, note that there is no point in stating 'it was recently reported by Strong and Brave (2014)...' when you are writing your paper in 2015. The cited work is dated as 2014, so it was obviously recent when you wrote your manuscript; however, when the same paper is read in 2035 it will be clear that the reported information is no longer recent.

Finally, remember that mistakes can be made in compiling a reference list. The most common errors are those of omission (i.e. a citation in the text is not in the reference list) and superfluous inclusion (i.e. an item in the reference list is not cited in the text). You may be guilty of both, so check your reference list carefully. Editors and reviewers can be very good at spotting such errors, but once again remember that authors are responsible for the accurate presentation of their papers. The advice is to check over the citations and reference list just before submission. It is a good plan to ask a willing colleague to help because others often spot errors that you have repeatedly overlooked through familiarity (which applies to the whole article as well).

8.8   The Digital Object Identifier

It was hoped that the *digital object identifier* (DOI) might provide a way of citing references in the text or at the end of a paper. The DOI is a coded number unique to each paper published in the literature, e.g. DOI: 10.1057/CBX2009066. This is in the form of an active link to literature search engines and databases, so the full reference need not be included in the paper, but this has not become common practice. The reader working online can double-click on the DOI and have the complete reference, or at least its title and abstract, on screen almost immediately. It is clear from today's behaviour that scientists would rather have good citations in the reference list than have to click on otherwise 'anonymous' DOIs to obtain a small piece of information needed at some point. When your paper is published it will be assigned a DOI, usually found in the footnote at the bottom of the first page.

We have now covered almost everything you need to prepare a good scientific paper. However, there are still some final matters, which will be dealt with in the next two chapters (9 and 10). The first will deal with figures, tables and illustrations, and the second will cover the critical reviewing of the 'final' manuscript, checking it with all the authors and preparing a letter to accompany the document prior to submission.

9   Figures and Tables

A research article is *the final product* of an investigation. It is a report that tells the world what you have done and found. On this account, it must be presented in the best possible way so that its message is to the point, clear and succinct.

9.1   Introduction

It is hard to generalize about figures and tables since they can be presented in many different ways, so only basic guidelines and general advice can be offered in this chapter. Their format must comply with the journal's requirements; however, some ubiquitous features might not appear in the Instructions to Authors. There is one almost universal rule: the figures and tables are not inserted into the text of a manuscript but placed after the References (though you might be asked to indicate in the text where each item might suitably be located). The publishers will set up your article to fit their page layout, usually a two-column arrangement. Online journals often have side panels to the text pages in which 'thumbnail' figures or tables appear near the relevant text. You click on these to open them up fully.

Space limitation influences not only where figures and tables are placed, but also their size in the final publication. During the last 20 years, figures have tended to become smaller (look at a *Nature* article to see just how much reduction is frequently used). This can lead to loss of essential detail, especially in images such as electron micrographs. Authors and editors should alert publishers to figures that need to be above a certain minimum size.

Two problems arise:

- how much detail to put into a figure or a table
- how much information to put into its legend or title, especially if details have been given in the Materials and Methods section.

Many writers find it difficult to get the right balance. The real issue is: how much is a figure or a table a 'stand-alone' item, i.e. can it be understood without too much reference to the text?

9.2   Figures

9.2.1   Graphs, Histograms/Bargraphs and Similar Items

These types of illustration not only convey data but give important information about their meaning – 'a picture is worth a thousand words'. Diagrams, schemata, graphs and histograms/bargraphs are usually more effective ways of communicating than merely tabulating the data or incorporating them into the text of the Results section. A reader (another expert in the field) might notice a potentially interesting article, read the Abstract and then glance over the illustrations. Such a reader would know what type of data to expect and how they might best be presented; this makes it easier for them to decide how much time to devote to the paper, i.e. its relevance and importance for them.

The amount of information that readers can assimilate quickly is limited. Judging how much data to put into a figure takes skill and experience (usually gained from giving PowerPoint presentations in front of your peers). As emphasized in previous chapters, it is wrong to select only your 'best' experimental data for a figure. There are some findings that carry little evidence that ought to be considered for selection to better balance the overall conclusions. 'Rogue' points that do not fit comfortably into the general pattern might be valid, although you cannot yet explain them, and they could mean something to a reader. Indeed, they may lead to refining or even replacing your hypothesis. All research is progressive. Each finding opens doors to new avenues of investigation, so a rogue point can sometimes indicate a new direction.

If you have many separate sets of data to graph, perhaps as many as 12, it might be best to segregate them into groups of, say, 4 per graph – remembering to use the control in each of them and not to alter the axis

(a)

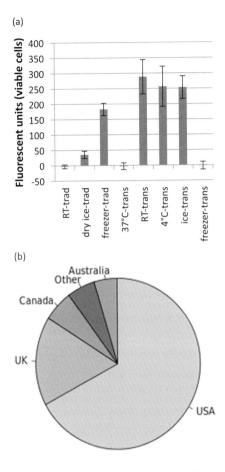

(b)

Figure 9.1 An example of (a) a histogram and (b) a pie chart. These two examples are shown to introduce their forms; figure construction is discussed in Section 9.2.4.

scales. This makes it easy for readers to make comparisons within the 12 data sets without being too confused. Sometimes it can be difficult to decide whether to use a *histogram* (see Figure 9.1a) or *line graph* (see Section 9.2.4 and Figure 9.2). As a rule of thumb, histograms have far fewer data points than line graphs. When the percentage of a whole is being depicted, a *pie chart* can be used (Figure 9.1b).

Under some circumstances you might want a figure to display the relationship between two variables – how they correlate – and then introduce, for example, time on a third axis. A 3D graph is hard to construct, but if you have enough experience to handle the data competently and have the right software, it can be worthwhile.

9.2.2   The Importance of Figures: the Art of Correlation

What are figures supposed to do, and how are they best constructed from the data? Much of science is about finding correlations. These can be classed as direct or indirect positive correlations, direct or indirect negative correlations, and cases where there is no correlation. Most experiments are carried out to determine whether correlations among the data support or refute a hypothesis (e.g. helping to confirm or deny a cause–effect relationship). Positive and negative direct correlations tell us that two variables are (probably) related in a proportional or an inverse way, respectively. The strength of a correlation can be checked by carrying out regression analysis, for which software is available. Regression coefficients – R values – can range from 0 to 1, the former being no correlation, the latter being a perfect correlation. Indirect correlations suggest that the relationship between the variables is coincidental, not causal. They can arise because (or where) an unconsidered third factor is involved, or by pure chance, e.g. when too few data points are available. This shows why all experiments need to be repeated, perhaps more than twice, so you are not misled by chance correlations. Repetition is also needed to confirm direct correlations, increasing confidence in the results. Good science always requires consolidation of the findings.

Repeated confirmation of new results gives the author confidence, but it does not guarantee acceptance by the rest of the scientific world, which is why independent corroboration (confirmation) is imperative. If other scientists obtain very similar results, consensus is attained and the new data become generally accepted. If others find quite different results, the experimental procedures, conditions and other factors have to be explored to find the reasons. This may resolve the difficulties one way or another, i.e. the hypothesis under test is in essence valid when the differences can be explained by the effects of mitigating factors, but it is quite possible that the findings cannot corroborate the evidence and the hypothesis might need to be rephrased or perhaps abandoned.

Correlations

Science is mostly about finding connections between one thing and another, i.e. *correlations*. Often – not always(!) – correlations can indicate cause and effect. For example, X is correlated with Y, causing a rise in temperature. But if I write that A is correlated with Y, this action might come via A acting on B, B acting on C, C on D, and so on until X finally acts on Y, i.e. a very loose and indirect effect indeed.

Science is *progressive*; a firm cause–effect relationship can lead to a new hypothesis that then needs to be tested. Science is therefore open-ended. If a logical sequence of correlations is found, this usually strengthens earlier hypotheses and can generate better understanding of a phenomenon, perhaps leading to a theory. When a hypothesis is strengthened in this way, the number of assumptions on which it is based can normally be reduced.

Negative evidence is seldom published separately, but it need not be omitted from a developing sequence of steps in a paper. As indicated earlier, there is good reason to include contrary evidence: omitting it could be detrimental to you and to others in the long run. Negative evidence is especially important where a hypothesis has become received wisdom (i.e. is generally accepted), but you have now called it into question, showing that it needs to be revised or replaced. In this case, corroboration of the negative evidence by independent researchers is essential. However, negative evidence that is not accompanied by new explanations for the phenomenon under scrutiny has less chance of being published.

9.2.3   Data Points

The current trend is to indicate that an experiment was repeated three times with similar results, represented in the figure by one of the replicates. The data points from all trials can also be pooled in a figure, but this practice seems to be dying out, although it has merits in some cases (especially when each of the separate trials involved small numbers).

The data encapsulated in a single point are usually the average of a set of values (e.g. when $a = 2$, b had values of 3.1, 4.6, 3.7, 5.2, 2.8. . ., which gives an average of ~3.9). Thus, means are calculated and plotted, but we also need to know the 'spread' or range of values about the means. These deviations can be estimated, giving confidence levels that allow a mean to be compared statistically with other means. Before you make such comparisons it is usually best to seek advice from an expert statistician, even when you think you can apply a suitable software package competently. (This is especially so where complicated comparisons are made between values found under different circumstances.) Standard errors (SE) or standard deviations (SD) should normally be included in the figure as bar lines above and/or below each point. The greater the SD, the greater the spread; but there is more to an SD than this, which is why it is worth consulting statisticians to ensure the best presentation of your data and to learn for yourself the true meanings of statistical values.

Correlations involving many data points can be established by *regression* analysis, and the line generated by an appropriate computer algorithm can be shown on the figure along with the regression value (1.0 being a perfect correlation between all the points, 0 being no correlation at all). There is also software to generate 'best-fit' curves for other sets of data points. It is up to you to decide what conclusion to draw, but remember it is dangerous to use any algorithm inappropriately.

9.2.4   Constructing Figures

Many books are devoted to good experimental design, and the preparation and analysis of data for best presentation, but here I will simply outline some of the ways in which figures, particularly graphs, can or should be constructed.

Suppose a is directly correlated with b, a being the independent variable; i.e. a is set by the experimenter (if it is the concentration of a drug, it might be increased by a factor of 10 in successive experimental groups). These a values are usually plotted on the *abscissa* (the horizontal axis). The mean values of the dependent variable, b, are plotted on the *ordinate* (vertical axis) with SE or SD indicated at each point. In each case the scale needs to be considered. In some graphs, both scales are linear. Scaling in other graphs may be linear-log, log-linear or log-log, or a and b could be related by their squares or cube roots. A negative correlation can be plotted as

$1/b$ instead of b. Correlations are often obvious, so little time is needed to plot the graph. If plotting proves more difficult, try different transformations of the data as suggested above. This exercise can tell you more about the data than was initially apparent, e.g. re-plotting on linear-log axes could reveal that b is an exponent of a.

One general feature of graphs is that the lowest values of a (if not 0) should be at the bottom left – the origin – of the graph. Values of b are also best plotted through 0, but sometimes we do not have zero values so we have to use some other baseline. We can then subtract this baseline from all the data and *plot their differences* from it, positive or negative. If the baseline values differ among groups we might also need to *normalize* them, i.e. make the data points for all groups equivalent (having the same baseline). For example, if groups 1 and 2 had started from baselines of 2 and 4, respectively, the corresponding a values would be adjusted as if both sets started from 2. This works easily for linearly related variables but is more difficult when the relationship is non-linear.

Another general piece of advice about drawing graphs is to make the ratio between the axes approximately 4:3 or 3:4. Excessively long or tall plots should be avoided. This is a matter of choosing the correct scales for the abscissa and ordinate. Poor graphs are often seen when, for example, the abscissa represents long time periods (months, years, or millennia) and stretches right across the page. The ordinate values may appear to differ only slightly, as illustrated in Figure 9.2a, in which case a change of scale can make the differences obvious (Figure 9.2b), these graphs showing the rainfall over the months of a year. The magnitude of the changes becomes much more obvious in the latter, indicating almost three times as much towards the end of the year, which is not easily appreciated in the former.

However, you should at all costs avoid misrepresenting your data. Potentially misleading representations can be illustrated by the movements of stock exchange share prices over a period of a week or a month. When the ordinate is scaled from just below the lowest share value to just above the peak at a particular time, the fluctuations look huge (Figure 9.3a). However, on an axis scaled from zero to the highest point they appear negligible (Figure 9.3b). Software programs choose minima and maxima in the way set out in Figure 9.3b, so check you have the correct scales.

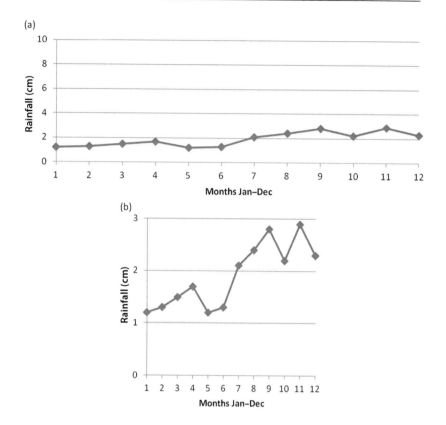

Figure 9.2 Line graphs. (a) A graph drawn with poor consideration for the ordinate scale (vertical axis) being too great, and the spread of months on the abscissa (horizontal axis) being too long. (b) By contracting both up, the differences in rainfall per month through a year become much more easily appreciated. In some cases the line graphs are preferred in a fully boxed format. Comparisons between different points can also be more easily estimated if gridlines are present, but these should not overwhelm the graph.

Axis ratio of figures

A ratio of 4:3 (or 3:4) should be used as far as possible in drawing graphs; avoid tall or long graphs. To do this, you will need to manipulate the scales of the axes carefully.

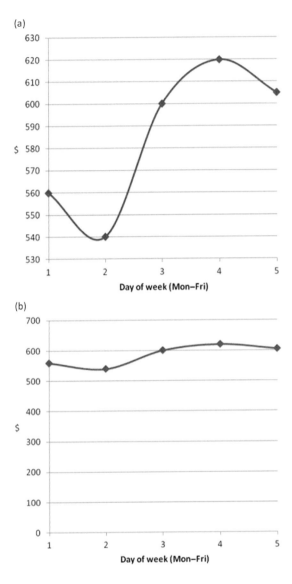

Figure 9.3 Data can be misrepresented by poor choice of scales. (a) Graph indicating change of the relative dollar value through a working week. Using an ordinate scale that covers quite closely the low point and the peak, a huge surge is seen between days 2 and 4. (b) Correctly plotted with the ordinate going through the origin (zero), the fluctuation *in real terms* is now seen as a small ripple. Exaggeration of this kind is most unwelcome in science, unlike the world of commerce!

Figures 9.2 and 9.3 are line graphs, but you can also represent your data as histograms, pie charts and in many other ways. It is best not to overload the figures with data. Readers want to see the main findings almost at a glance.

Your attention should be drawn to the ordinates (vertical axes) in these figures. In Figure 9.2a, the scale from 0 to 10 is pointless when no value goes over 3. Fluctuations can be seen much more clearly when the axis is from 0 to 3. See what happens in Figure 9.3a and 9.3b, however, when only part of an axis is used instead of from a base of 0. If the market price of something like gold is being shown, it would seem it was on a roller coaster according to Figure 9.3a, but not from Figure 9.3b. It is important therefore how you present data, as it can be quite misleading if presented badly.

This manual does not use colour except on the cover. Colour is often used to enhance the appearance of line graphs, bargraphs and pie charts. Now that most articles are online, publishing in colour is no longer an expensive alternative to black and white. This was not the case when hard copy was the only choice for publishing papers before the electronic era arrived. It may seem easy to present lots of information using differently coloured curves, but there is no need to overuse this facility. It is also best to steer clear of weak hues because they do not stand out against a white background, e.g. yellow is always best avoided. In Figure 9.4 we can see that when only black and white are used the possibilities of representation of each curve are quite numerous. Some journals specify a preferred order: e.g. open circles, closed circles, open and closed squares, open and closed triangles, etc. Lines can be solid, small dots, short lines (dashes), long dashes and even a mixture of dots and dashes, which gives considerable versatility (Figure 9.4). This can even be extended to varying the thickness of the lines being used. When there are a lot of data to present, having two or more separate graphs, each showing the common control, makes the data much easier to comprehend.

9.2.5   Figure Legends

A figure should have enough information in its legend to make it clear what it represents, i.e. a reader should be able to grasp its message without referring to the text (the figure is essentially self-explanatory). However,

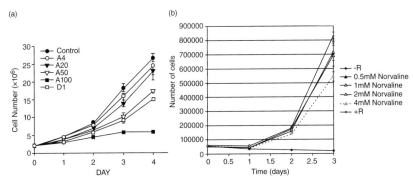

Figure 9.4 (a) Graph of six groups giving mean values ±1 standard deviation as bars around the averages. Bar lines can be so close that they disappear into the symbol for the average value. In this graph the symbols are diamonds, inverted triangles and squares, filled and unfilled, allowing six curves to be plotted. Any more than this would make it difficult to untangle the curves. (b) A similar graph with gridlines included, which can help in assessing how much difference there can be between the different curves. Here, the small symbols and weak bar lines for the standard deviations do not help the reader. The use of a dotted line does help to distinguish a curve in this case where treatment with the agent seems to be becoming effective. Note that in both graphs the treatments that were used are included, within the graph in (a) and to the right-hand side in (b). The alternative is not to put this information in the graph itself, but in the legend.

there is a growing tendency to put unnecessary information into a figure legend, including details that reiterate information about the experimental design and methodology, for which the Materials and Methods ought to be consulted. The title of a figure must therefore indicate the part of the evidence for (or against) a hypothesis it presents. It should be short and to the point. If the curves, blocks in histograms and so on are not labelled on the figure itself or as a side box or list (as in Figure 9.4), they need to be clearly specified in the legend. Any unusual features should be mentioned. If a figure represents several similar (identical) experiments, or pooled data from several trials as mentioned before, the number should be specified. Appropriate statistical treatment of the data to indicate the probability that differences exist (P values) between control and experimental groups can also be added on the figures themselves or included in the legend, although they will also be covered in the main text.

9.3   Tables

9.3.1   General Remarks

Tables are easier to construct than figures. They can contain little information or a vast amount. It is difficult to generalize, but if a table looks extremely long, consider constructing several separate tables instead.

Tables need titles that tell the reader their exact relevance to your paper. Making the cells in a table much bigger than the information they contain is a waste of space. However, your readers do not usually carry magnifying glasses around with them when they read hard copy; even zooming in on the computer can be a problem because often only part of the table is enlarged, making it difficult to compare it with the rest. The advice is not to use too small a font, and try landscape if portrait format looks too cramped.

It can be difficult to decide what information should go into the vertical 'axes' (columns) as opposed to the horizontal 'axes' (rows). Usually the known parameters or groups under examination go in the former and the values or data in the latter. If mean numbers are inserted, ± SDs or SEs can go alongside them or in adjacent columns. Probability values can also be inserted. A column on the right-hand side might be used to give references if the data are compiled from your own results and other people's.

Special notes about the information in any of the cells can be appended in small type below a table using superscript symbols in the cells concerned. (Sometimes information about the methods used and these other special notes are put immediately after the title, but my advice is to have the title on its own, clearly stating what the table is about.)

One important point about tables must be emphasized. The information they contain can also be recorded in the Materials and Methods, Results and Discussion. Much of the information in a small table could perhaps be given in the text of the Results, but avoid repetition. Some people append information from the Materials and Methods section in the footnotes, which is usually unnecessary. Try to present all your information in the appropriate places, with maximum clarity, but without unnecessary duplication. What ought to be avoided is to have the Results section full of data, such as values for 20 items within 4 categories, that would be much easier to assimilate if tabulated.

9.3.2   A Simple Example of a Table

The box here contains an example of a small table. I am going to draw attention here, as I could have done when discussing figures, to the numerical values of the data. Many authors cannot distinguish between precision and accuracy.

Table 1: Effect of *x* on blood glucose levels in mice

Exp group	Dose of *x* (mg)	*n*	% of control	Blood glucose (mg/dl)*	SD
A	10	17**	28	7.3	1.423
B	1	18	91	7.7	1.672
C (control)	0	18	100	7.81	1.233

* on day 7
** 1 died day 6

Note that I have not used gridlines – the table is easy to read without them. If the amount of information makes it difficult to follow which line is which, then gridlines can be included. The information about what was done is self-evident, so there is virtually nothing to add about the Materials and Methods, other than to indicate by an asterisk that blood glucose values were recorded on day 7 (it is quickly inferred that this means seven days after treatment began).

9.3.3   Accuracy and Precision

What is said here applies to other places where numbers are used – in the text, figures, etc. I have deliberately given a control blood glucose value of 7.81 mg/dl and also stated SD values to three decimal places. The other values are sensible because they are sufficiently accurate. Accuracy in the values in such a table reflects how exactly blood glucose can be measured. Let us say you use a meter for the measurements. You should first check the precision of the instrument. Suppose the manufacturer states that model BLOGLO-Mark2 is a 'precision instrument'; so when it says it is delivering 10 ml, you expect it to do exactly that. But precision is relative. The meter may be reliable to 0.01 ml, which is indeed very precise. However, under your test conditions against a standard, you might get a relatively consistent reading of 9.8, but when you have taken this reading 10 times the range is from 9.7 to 10.3, a variability of about 6 per cent. So your instrument lacks the expected precision. This makes it possible to average your 18 control readings, as in Table 9.1, to no better than 7.3–7.7. A mean value of 7.81 is therefore too precise; 7.8 is more meaningful, i.e. it is as *accurate* as your instrument permits. Similarly, the SD can, if necessary, be calculated to many decimal places, but accuracy in this case requires only one decimal place – using three decimal places, as in the table, is inappropriate. They can be rounded up or down in the usual manner.

9.4   Schemata, Diagrams, Line Drawings, Symbols and Units

All other types of illustration need to be kept 'manageable' for the sake of the reader – not too much information; even if they are quite complicated the points they make should be clear. Remember that the publisher will probably reduce the overall size considerably, so avoid small fonts or little pictures and symbols, or the published copy will be difficult to read. In online illustrations, the size can be altered by zooming in on them, which obviates this problem. Schemata and diagrams need to be as self-explanatory as possible, so it is best to use very short titles. When more explanation is needed it is best to refer back to the text, despite what has been said about legends to figures (graphs). The use of colour helps, and today there is little restriction on its use.

Line drawings ought to be as clear as possible with neat clean lines. Titles should be short; line drawings also need to be self-explanatory. Since they may be reduced in size, do not make them too faint or the lines too thin; grey is quickly lost on reduction (degradation).

Images, photographs, microphotographs and similar illustrations can present the following problems:

(1) The file size may be too large if top quality is presented, making compression necessary.

(2) The pixel density may be too low – a compromise with (1) is needed.

(3) The contrast must be good.

(4) The brightness (e.g. in fluorescent micrographs) must be adjusted for features to stand out, not least when images are merged.

(5) A scale is usually needed, which is best put on the images as bars rather than stated as magnification in the legend – magnification is always affected by the reduction of the illustration during production. Try to use scientifically approved units (e.g. μm). Since a scale must have units, it is important to be accurate and consistent with their use, and with the symbols often associated with them (see the Further Reading section at the end of the book).

(6) Proportions may vary more than in graphs, depending on the subject depicted, but 4:3 or 3:4 ratios are still best if possible.

(7) Labels need to be added, a good example being the use of, say, *n* placed on a nucleus of a cell, and the meaning of the abbreviation explained in the legend.

Fluorescent images can be problematical because they often have black backgrounds against which the fluorescent objects, often quite dim, must be visible. During preparation for publication, these images go through several steps; at each the quality of the image becomes degraded, and contrast is lost. This means that high resolution should be used when the images are taken, and adjusted to give brightness and strong contrast. One particular reason for this is that several images often have to be compared side-by-side. When printing in colour is expensive, e.g. in hard copies of articles, publishers make these images very small (thumbnail size), which greatly exacerbates the problem. Today, electronic publication circumvents these difficulties, but it is important to keep these three main considerations in mind: high resolution, brightness and strong contrast.

Other issues arise with the need for fine detail in photomicrographs, notably electron micrographs (EM), where reduction in the size of the image spoils the resolution. The publisher may reduce the size of images to fit the page, or column width, without necessarily appreciating the author's requirements. Thus, where this is important, the publisher should be informed that the EM should be kept at their full size (again, this is less relevant to online publication).

One final warning: it is all too easy with modern software to manipulate images, to 'retouch' photographs or other illustrations. Where comparisons are being made, altering one of (say) a pair but not the other can make a huge difference (e.g. in fluorescent images and Western blots). There are no general rules – except you must not alter an image you wish to include in your article in order to prove a point that is not valid. Modern software can detect airbrushing and similar manipulations that make data suspicious, so if you broke this rule you would be found out.

9.5   Supplementary Material

Adding supplementary data to an article is an excellent way of providing information that supports a claim in the main text but is not crucial to the hypothesis being tested. This 'section' can contain material in any form (text, figures, tables, etc.). It has become increasingly common in online publications. Before the advent of electronic publishing, space was a major consideration, but supplementary material (and indeed a whole article) can be extremely long nowadays. The problem with long publications is that they try the reader's patience, so it is still advisable not to overload a paper with information.

10   Presubmission

A research article is *the final product* of an investigation. It is a report that tells the world what you have done and found. For this reason, it must be presented in the best possible way so that its message is to the point, clear and succinct.

10.1   Introduction

You might suppose that everything about writing a paper has been covered by the end of Chapter 9. However, some additional tasks must be done before you submit your manuscript:

- Rereading of the *Instruction to Authors* of the first choice of journal for submission
- Revision of the entire manuscript by the main author.
- Final revision to be undertaken after all the other authors have (re)read the manuscript and commented on it.
- Confirmation that all authors have agreed to its submission.
- Checking whether any of your co-authors have conflicts of interest.
- Presentation of the article in the appropriate language, usually in English; for non-native speakers, find a colleague who is a native English speaker to help, and/or get help from a professional editorial service company (see Section 10.4).
- Ensuring that you have informed and/or shown your head of department/line managers/authorities that you have a paper ready for submission (mandatory in some institutions, e.g. the National Institutes of Health in the USA).

- Checking whether the different sections (and subsections) of the paper need to be numbered and in the required format, as specified by the journal to which it is to be submitted.
- Establishing that all the necessary files – text, figures, tables, supplementary material, etc., are at hand, ready for uploading into the journal's submission system.
- Preparing a letter of submission that will go with the article to the chosen journal.

Some of these points are self-evident so I will not spend much time on them. Some might not be relevant to the manuscript on which you are working, but it is still worth keeping them in mind that you will need to consider them again when you write your next articles.

10.2   Revision by the Main Author

It is unlikely that a paper in parts can be written by a group of authors and can later be simply assembled; one person normally carries the task of preparing the whole manuscript. When you have gone as far as you can in this drafting, it is best to have a few days' rest to take your mind off it (writing a paper can be difficult, tedious and tiring!). Returning to it afresh after several days often results in some surprising findings (e.g. simple mistakes such as missing a space or a symbol, finding a phrase that is incomplete, repeated or ambiguous). You might see that a figure poorly represents what you are trying to show or a citation is wrongly located in the article. When the changes have been made, the paper might be reread several times as if you were the *reader*, not the writer, and the fine honing will soon be completed. This process can be never-ending, so there comes a time when you have to stop, but you will now need the approval of your co-authors.

10.3   Revision by Co-Authors

Your final version is given to your co-author(s), who should go through it as they will see it through different eyes and might still detect something that needs to be changed (e.g. misspelling of an author's name in a citation). The final touches are then made ready for submission. Unfortunately many co-authors put little effort into checking a manuscript, so there may be some small corrections to make, and this will not be much of a chore. If this is the case, at least the co-authors have seen the paper and have no grounds for redress if later they take issue with some of its content.

Revision of a manuscript

You cannot make too many revisions of a manuscript; almost every new attempt will lead to improvement, but you have to draw the line somewhere. You will also get criticisms of your paper from your co-authors or close colleagues, and from the referees after submission, so the process will continue until the article is finally appraised by the editor.

A statement that approval of a paper was given by all the authors can be placed in the acknowledgements or as a footnote on the title page. The two stages covered here and in Section 10.2 usually result in a much improved article, which makes the effort worthwhile.

10.4 English Presentation

Papers in international journals are almost exclusively published in English. If you or your close colleagues do not have a good command of English grammar and idiom, it is best to send the final draft to a *professional editing service* that will do this job for you. Unfortunately, in many of these services, the staff may be neither native English speakers nor adequately qualified in the subject area of your article, so choose carefully. While there are charges for these services, the best of them can greatly enhance your chances of an article being accepted by a journal, provided the science is original and good.

10.5 Author Sequence and Affiliations

The order in which authors are placed below the title of a paper is a matter that has to be mutually agreed and is discussed in Chapter 8. (However, as a last check before submission, it is worth going over many of these issues again, particularly if you are reading the chapters of this manual separately as you go through the stages in the production of a manuscript.) This is important because there can be much dissent on occasions as to the best arrangement. Sometimes the head of the laboratory, often the principal investigator, will want to come first, although many of those who are well established and have published several previous papers on the research will decide to go last. Some heads of department insist that they are included in the author list usually because the work has emanated from their unit. The question of whether someone, e.g. a postgraduate student, has significantly contributed and should be a co-author is tenuous, depending on circumstances. If they have contributed little more than occasional technical help, they will probably not be included, but it is debatable whether a technician who has contributed significantly to the project ought to be listed; the alternative here is to record that person's contribution in the Acknowledgements.

The affiliations of each author need to be noted under their names below the title. When an author has moved away from the place at which the work was done, their new affiliation must be recorded. In Chapter 16 on ethics, we will find reference to cases in which an author has moved away from the laboratory where the work was done and then published the findings as if they were their own, leaving out all their colleagues who should have been co-authors. This can lead to unpleasant repercussions and must be avoided at all cost.

The corresponding author (not necessarily the senior author) has to have an email address included; some journals insist that the emails of all the authors must be given.

10.6   Conflicts of Interest of the Co-Authors

Although detailed before, in Chapter 8, remember here that all conflicts of interest have to be included. These are mostly cases where work has been done in a laboratory that has been funded by some company which has been paying one of the authors to be involved in the project. However, as the senior author, you will know of your own conflicts of interest, if any, but perhaps not those of your co-authors, who must be approached on this matter. If a case is obvious to you, it is worth checking with each co-author as to any possible conflict of interest that might have to be noted and explained. Ensure that conflicts of interest are recorded in the appropriate section of the paper, which follows the Acknowledgements.

10.7   Sections and Subheadings

Some journals have numbered sections (1, 2...), with subheadings given subsets of numbers (2.1, 2.2... 2.1.1, 2.1.2... and so on). For illustration, look at the subsection numbering in this chapter or earlier ones. Major section headings might have larger font, be in bold type, use a title format or be written in upper case. Try to comply with the journal format in respect of these details, though the publishing editor will make sure the final presentation is correct.

One section of a paper often lacks subheadings, i.e. the Discussion. This part of many manuscripts can cover many pages, but as explained in earlier chapters it is best to be as succinct as possible, developing short arguments on four or five major points. This makes it relatively obvious where appropriate subheadings should be placed. Without them, the reader can find a Discussion tedious and difficult to assimilate.

10.8   Letter of Submission

In days gone by it was expected that the journal editor would receive a letter from the authors explaining the purport of the paper and why they believed the chosen journal was suitable (e.g. because it has already published many good papers on the same subject). The letter might also include some of the matters raised above. Today, most articles are submitted online, but some journals still prefer to have some information similar to that of a covering letter. Where some important information on the nature or content of a paper ought to be explained by the author to an editor, this should definitely be included at the submission stage; alternatively the editor can be contacted directly by email.

Pre-checking a manuscript

If there is any doubt that the journal of your choice is a suitable venue for publication, you can pre-check it with the editor by sending an email along with the Abstract of the paper (or the whole manuscript); see the following text.

In addition, it is often useful to send an email to the editor to pre-check the suitability of your article for a particular journal. Many authors send just the Abstract as an attachment. As an editor, I sometimes find that the Abstract is insufficient to judge the suitability of a paper, in which case I request the whole article. An Abstract alone seldom indicates whether the paper is short and to the point or long and rambling. The advice therefore is to send the whole article – it involves no more effort.

An email is also needed if particular issues or circumstances need to be intimated to the editor, for example if the paper has been rejected by a higher impact journal but not because the work is unsound, or if it is an important corroboration of other reports and therefore not necessarily highly original or novel. (A host of other reasons might be added, but these examples will suffice.) This procedure might help to establish whether a paper of seemingly 'marginal' interest could nevertheless be welcome. In many cases it prevents authors wasting time trying to get an article published in the 'wrong' place. When the authors and editor come to an agreement that the manuscript is worth submitting, this does not mean that it will have an easier passage through the editorial stages since the referees, unaware of any dialogue between the author and editor, will now look very critically at the science involved and decide whether or not it is acceptable, and then the editor and editorial board will take the final decision after weighing up the pro and cons from the referees' reports.

10.9   Concluding Remarks

After a long and exacting process, we have now reached the point where your article is ready to be submitted. Its journey thereafter will be covered in the following chapters, but from my own experience I have found that authors are rarely sufficiently familiar with what is involved in the submission process, and are even less familiar with the editorial and publishing processes that follow. This is important to know as it helps to smooth out many of the difficulties met after the article has been sent in by the author.

There is one last word on the subject of preparing and submitting a paper for publication. This involves ethics, which will be discussed in more detail towards the end of this book, notably Chapter 16. However, some key points ought to be noted here. Remember that your paper should not duplicate any previous publication of your own or resemble anyone else's work. Small passages of text or a figure from other papers can be included if permission is given and these instances are acknowledged in the text. In brief, your work must not be plagiarized, it must be original. The article cannot be submitted to several journals at the same time; it can only be submitted to them in succession following rejection from each one. Multiple submission is *highly unethical* and can cause a lot of unnecessary work for editors and publishers. Fortunately this practice is now being 'policed' ever more assiduously, thanks to the introduction of increasingly sophisticated software that detects such unethical practices.

Nota bene: Some information, instructions and advice in this manual will be seen as repetitive. I am aware that this is not normally good practice, but it is *not unintentional.* Readers of a manual looking for guidance delve into different chapters/sections for specific information at different times. A manual is consulted, not read from cover to cover. Repetition in these circumstances will not be so self-evident; indeed, it can help to emphasize and reinforce many different points during the learning process.

11 Submission of Manuscripts

There is a general consensus among editors and publishers that it is the author's responsibility to submit a paper in accordance with the format of the journal to which it is submitted. Take heed!

11.1 Introduction

The practicalities of submitting a paper to a journal run by a publishing house depend on your preparedness, which has been explained in Chapter 10. It can be a lengthy process to complete a manuscript ready for submission and the advice given here should be taken seriously. Many will submit a paper that is a long way from being fully prepared for submission and get annoyed when the editor of a journal sends it back for immediate changes to comply with a journal's requirements. To avoid or reduce this frustration, ensure you follow the directions in Section 11.2.

Another problem frequently met by authors is that they might be sending a manuscript out to a second or third journal, having already been rejected by one or more editors. This is frustrating, but the paper has to be revised in accordance with the requirements of each journal that has been tried. Editors will notice very quickly if a paper has been prepared for another journal before theirs when appropriate changes have not been made. This exercise has some advantages because a rejected paper may have been peer reviewed several times, making it possible to meet the criticisms and thereby improve its content as well as the formatting. The author has gained much from this interaction and should be extremely grateful for the valuable feedback.

The submission systems used by most publishers are nearly all electronic with online access, many of them using similar software. Some are complicated and take some time to complete, as will be seen below. Your task will be to submit to the most appropriate journal, remembering that the chances of acceptance for a relatively average paper will be small if submitted to a high impact factor journal. Therefore make your choice at an appropriate level and decide whether your paper ought to go to a *general* journal or a *specialist* journal in the field of your topic.

If there is any doubt about which journal to choose, look at papers similar to your own that have been published in the last 2–3 years and check which ones were selected. To be even more sure that you have made the right choice, send an email to the editor of a journal with the abstract of your paper (or indeed the whole paper as an attachment) asking whether it is suitable. If the answer is yes, you will be directed to the website for the submission process. These measures can often save a lot of time and frustration at the start of the submission process, especially where there would be a triage stage that is tantamount to getting feedback from the editor on the suitability of your paper.

11.2   Complying with Journal Requirements

There are three main considerations in starting the submission process:

- You must follow the guidelines, the Instructions to Authors, usually specified rather precisely by the journal, as to how your paper should be formatted ready for review. If you do not comply, the editors may keep referring the paper back to you until you do comply.
- You must also be familiar and comply with any guidelines that come from the publisher.
- If you use British English, it must be consistent throughout the paper; the same applies for International and US English, i.e. do not switch from one to another in the same paper. The same also applies to SI units, acronyms and similar items, one of which is the referencing system, the Harvard style (Jones and Li, 2012) or the numbered Vancouver style (reference,[3] or reference [3]). You cannot change from one system to the other. All this has been fully explained in Chapter 8.

11.3   The Checklist

Thoroughness is the important thing to keep in mind. Chapter 10 has the details you require, but the main points will be reiterated here since you may be referring to this chapter separately rather than looking back at the previous one. You need to:

- Make a complete revision of the entire manuscript several times over, ensuring that grammatical and spelling errors are corrected, and that there is consistency in terms, units, abbreviations and other matters already discussed. If you have defined an abbreviation in the text (e.g. blood pressure – BP), continue to use BP thereafter and not the full words or a mixture of these and the abbreviation, i.e. be consistent.
- Ensure that the correct referencing system has been used (Harvard or Vancouver style).
- Ensure that your punctuation and letter and word spacing are correct. Too often the following is seen: *et al*, instead of *et al.*,. Another example is *treatment(Jones and Li)*, which should read *treatment (Jones and Li)*, i.e. with a space before the first bracket. Copy editors get annoyed when there are numerous small typing errors present; the author should have picked up on these by noticing indications in their software that mistakes have been made (e.g. red or green underline in Word). The same goes for spelling mistakes.
- Check whether the different sections (and subsections) of the paper need to be numbered and are in the required format, as specified by the journal to which it is to be submitted.
- Give your latest version to your co-authors, who may well find other errors or ask for small revisions, and then revise it accordingly to produce the final version.
- Get confirmation from all the co-authors that they are satisfied with the final version, and add their email addresses if the journal to be approached needs them.
- Indicate who the corresponding author is and give your email address and preferably telephone (sometimes also fax) number.

- Check whether any of the authors has a conflict of interest.
- Check that you have made a statement in the paper regarding the ethical permission when carrying out animal and human experiments (or surveys for the latter, as in research into healthcare issues).
- Ensure that you have informed and/or shown your head of department/ line manager/authorities that you have a paper ready for submission. Where necessary, get their permission before proceeding.
- Prepare a letter of submission that will eventually go with the article to the chosen journal.

Too many authors have not gone through a checklist before submission; doing so can save a lot of time. Some publishers provide a checklist at submission.

11.4   Charges Relating to Submissions

There are basically two main options for authors regarding the type of journal chosen. There are those that run an 'author pays' system, the other being usually without charge where the readership (the individual, library, organization, company or institution) takes out a *subscription* to have access online to a journal (often a basket of similar journals from the same publisher) or have a hard copy sent. Make sure you know that you are going to the right type, ensuring by whom and how the 'article processing charge' for the author pays system will be covered. In applying for research grants, it is best to include a reasonable figure in the proposed budget to cover publication costs.

Another problem, which is less restricting today than in the past, is whether you will need to pay for colour figures, tables, schemas, etc. This is seldom if ever the case with online journals as it makes no difference whether colour is used or not. Many publishers of hard-copy journals did charge quite substantially for colour images in the past, but they became less competitive when they did so, and therefore now accept papers with colour in them with little or no charge as long as the number of images is not excessive. However, it is wise to check beforehand whether you will need to pay, as charges can be quite substantial.

Charges for publication

Never forget that publishing papers can be a costly business. Always check who has to pay – author or subscriber/reader (often the reader's institution).

11.5   A Published Article as a Journal Guideline

Arguably the quickest way to check how a paper should comply with a journal's guidelines is to find and download a copy of one from the chosen journal from the Internet. (An example was given in Chapter 2.) It will show you how your paper will finally appear, although your typescript will not have precisely the arrangement found in the final published form (discussed below). A paper sent to a particular journal will be expected to have their familiar main sections, usually in the same order that we discussed in earlier chapters, although each journal has its own 'style'.

The publisher specifies guidelines for the *type of files* that can be uploaded to their server. They may also restrict the size of some files due to limited space on their servers, where space is taken up by all manuscripts and their updated and reviewing versions. These versions can be tracked during the reviewing process, and after publication online, as detailed in the next section. Figures or pictures of very high resolution requiring a lot of space should be compressed into a much smaller size. Note that a submission is uploaded to the publisher's server, making it immediately accessible to the journal editing staff, which are often independent from the publisher; for example, some editorial offices are run independently by an organization such as a learned society that has teamed up with a particular publisher. They will pass on acceptable papers close to the final format that has been requested by the publishing company.

11.6   Normal Formatting of Files for Submission

- Main manuscript file – Word Document (Microsoft Office Word 2003 to 2013, Office 365) or a similar word processing file for Mac computers. A PDF (Adobe) can also be sent, but the submission system will usually create it anyway. A PDF file allows one to see the paper very easily, but it is currently quite difficult to do a lot of editing on it, and that is why doc/docx files are needed.

• Figures and tables – JPEG or TIFF files (PowerPoint, BMP and PNG files may also be accepted).

• References – are usually inserted in your paper from Word Documents, RefWorks, EndNote or some similar software. These programs allow you to select and readily cite the relevant references in your article from an extensive compilation you have made during your research work. The software provides templates compatible with the journal style (or one very close to it), into which you can upload the selected references.

• Other supporting documents – sometimes papers can be enhanced with multimedia adjuncts, such as videos or 3D images, which can be included in the published paper when submitting to an online journal, or added for clarity for the reviewers. Current accepted files include: AVI, SWF, MPEG, QT and MOV for films, PDB for structures and Flash for animated schemes. Check with the editor if there is any concern about the type of file, its size, etc., especially when trying to include video clips.

All the files that you need for a paper should be ready before they are uploaded into the submission system. Since the text, figures, tables, supplementary files, etc., are going to be dealt with separately during preparation for publication, you should not upload a single file containing the whole manuscript (MS; except for a PDF version, if you wish to include it – though this is usually automatically generated from the files that you submit). The text file is followed by separate files, as mentioned above. You can indicate in the text where you would prefer to have figures and tables inserted, although some journals use side panels in their online papers, from which the appropriate figures can be enlarged, at about the right place in the text.

11.7   Submitting Your Manuscript: a Quick Revision

The publisher may specify the type of files that can be uploaded to their server, so check that the ones suggested here will comply. They may also restrict the size of the files due to limited space on their server (where all manuscripts and future versions have to be held during the reviewing process, and possibly after publication). All the files, apart from supporting documents, will be combined into the PDF file that is generated by the system itself on completion of the submission procedure.

Once the manuscript is in the appropriate format, submission can begin, but here is a list of files that should be ready before you get started, as previously mentioned (the checklist in Section 11.3):

(1) Main file of manuscript should include:
   (a) Complete text: title, affiliations (see point 1b), keywords, abbreviations, corresponding author details (see point 1c), a short running title, conflict of interest statement, Abstract, Introduction, Materials and Methods (sometimes put at the end of the text), Results, Discussion (sometimes the last two are combined), Conclusion (not necessarily a separate section), Acknowledgements and Funding, References, table and figure titles and legends (but not the figures or tables themselves). Each of these may need to be copied into the appropriate boxes in the submission system, which will keep you right about the order (which may not necessarily be the above).
   (b) Full names and affiliations of authors (no need to include all email addresses other than that of the corresponding author unless specifically requested).
   (c) Below the affiliations given in (1b), a paragraph should give the full name, physical address, telephone and fax numbers, and email address of the corresponding author (all correspondence and proofs will be sent to this author).
(2) Figures and Tables as separate files.
(3) Supplementary material.
(4) You may be asked to provide a word count.
(5) A covering letter (where appropriate).
(6) In some cases you might be asked to suggest potential (independent) reviewers, and you might also add unsuitable reviewers, e.g. where a conflict of interest might have arisen with third parties. Editors very occasionally fall back on nominated reviewers, but prefer to use their own choice to ensure truly independent evaluation.

11.8   Uploading Your Manuscript

Note: these steps may not be precisely the same for different journals because they do not use the same software programs; however, most are very similar.

Step 1 Most journal submission systems (JSS) will start with the creation of a personalized password and username. Open an account at the online submission site of the journal. This will usually require your name, email, affiliation, password of choice and further contact details. This will give you the access you need to your article during the review process post-submission (see below). It is important to make a note of these, so that any further revisions can be uploaded to the same place.

Step 2 Log into your account and click on the 'submit a new article' button.

Step 3 Follow the online instructions:

o Copy your title into the title box (sometimes journals limit the number of characters used).

o Copy the abstract into the appropriate box (again the number of words might be restricted). It is possible that you will be asked to upload the title and abstract separately. This can be done by having the manuscript open on the desktop screen so that the relevant information/sections can be cut and pasted.

o Choose the type of article – research article, hypothesis, review, commentary, etc.

o Enter any other authors' names, affiliations and emails. Note that most journals require each author's email address, each of which must be different and not for example a generic departmental email.

o There will be several questions asked about the manuscript, such as verification of its originality, and ethical matters such as ensuring you have approval for work done on animals or humans. There is likely to be a declaration where you will be asked to verify that the manuscript has not been submitted elsewhere.

o Enter the order of authors and co-authors that you require (the corresponding author submitting the article may not necessarily be the first or senior author).

○ Most systems will ask for a cover letter. This can be used to give an outline of your work, or any relevant information that you would want to give to the editorial office. As already mentioned, this should be ready before upload for convenience, and can be copied and pasted.

○ Upload all the files separately, main document (text), figures, tables and supporting material.

Step 4 In most JSS, the Word doc. file will be converted to PDF automatically (with figures at the end) so the editors and reviewers can see the whole paper in formatted style (a PDF file will keep the formatting and paper looking as submitted, rather than the reviewer viewing the paper through other programs that might otherwise change the look of the paper).

Step 5 Review the PDF for completeness and accuracy, and if correct, confirmation will probably be required to complete the submission process.

Journals today have their own submission sites online and prefer authors to submit using their specific formats. For example, the ScholarOne™ site (https://bit.ly/2YBiJX0) is used by Nature Publishing Group journals, many Cambridge journals, and is probably the most widely used system. Editorial Manager® is used by Springer, Wiley and PLOS, and other publishers will have their own systems. Elsevier's own proprietary system is called Evise. The problem is not one of finding it difficult to conform to the requirements, but of a paper being repeatedly submitted to one journal after another in a bid to gain acceptance. In short, since submission formats differ from journal to journal, authors can easily become frustrated and annoyed by constantly having to redo their submission formalities.

However, there are obvious advantages to using online submission systems, which include the following: (i) faster and more convenient entry into the publication process, (ii) saving everyone from authors to publishers both time and money, (iii) submission can be done from anywhere via the Internet, (iv) the submission site is accessible to only the involved parties (authors, editors, publishers) and (iv) the system makes it easier to move manuscripts on to reviewers and editors, thereby speeding up a final decision on acceptability.

11.9   Problems in Submitting Papers

For authors who are native English speakers, following the instructions to upload an article can often be difficult and frustrating, and therefore the task can be even more difficult for non-native English speakers. Indeed, it is quite usual to have authors complaining to editors and publishers that they have met problems in uploading papers. Although sending in a paper as an email attachment to the editorial office to have it uploaded by its staff is one solution, these offices would be inundated with papers if every author did the same, making the whole object of online submission pointless. However, editorial staff will usually guide people through these difficulties where, for example, Internet access and language problems seem quite insurmountable.

11.10   Dealing with Copyright at Submission

There are certain software submission programs that do not allow the editor or his/her staff to upload a manuscript on behalf of authors. This is because of embedded copyright and declaration issues in the software – for example, the author is the only person who can verify certain questions, such as certifying that the manuscript has not been submitted to any other journals at the same time. There is also the issue of using other people's text, data, figures, etc., which will require permission from the publishers or authors of the original material, otherwise copyright may be breached. Acknowledgements to these people for the permission to use the copied material must be made within the submitted manuscript in each and every case, otherwise you could still be guilty of plagiarism.

11.11   Preprints and the Future of Research Communications

'Preprints' – the publication of final drafts of papers without peer review – have been a practice in the physical sciences for several decades, but have only been increasing in the life sciences in the last decade, covering about 1–2 per cent of the total annual number of publications at the present rate (2019). Many reasons are put forward for their publication, but it raises several issues seen by some authorities as generally unacceptable, and by others as the way ahead. Since all science is communication, what a preprint represents is a state-of-the-art disclosure of work that has reached a point where it seems to be ready for communication to the world. Platforms for publishing preprints (also known as eprints) have been made available through several world-renowned institutes and organizations, for example Cold Spring Harbor Laboratory (which launched **bioRxiv** in 2013; www.biorxiv.org). This emerged after an earlier attempt by the National Institutes of Health (Bethesda, MD, USA) in the 1990s, that in turn gave way in 1991 to the open-access repository for physics called **arXiv**, which is now hosted by Cornell University (note: preprints were published on a non-profit basis; https://arxiv.org). Other attempts have come and gone, including one by Nature Publishing (called **Nature Precedings**) started in 2007, but this collapsed within 5 years.

Despite these trials and tribulations, in all probability the preprint is here to stay and is likely to become a regular mode of scientific communication, preferred by some but not all researchers. The more endorsements that the preprint system gets from major publishers, the more likely it is to become an integral pathway to publishing new findings. The major objection has to be that papers can be published without the proper scrutiny, i.e. peer review. It can therefore also be seen as a way of getting out some poor data, selected and inconsistent information, half-baked notions and so on for everyone to see. To be of value, authors have to be honest and scrupulous. Reviewing the fate of many preprints, it is interesting to note that some authorities have indicated that most have ended up as full primary research papers published in reputable conventional journals from which they will go into the literature repositories (available on PubMed and Medline), and will take on the impact factor of the learned journal in question.

I liken this situation to the debates in the late 1990s that led finally in the early 2000s to electronic publishing becoming the mode. A lot stems on not just honesty, but money. In Chapter 16 I refer to the 'author pays' model, and point out that BioMed Central, one of the first major electronic publishing companies, might have adopted the policy of its founder that it would publish everything submitted and leave it up to the readers that had free access to all the papers to decide what was good and what was trash. Fortunately this did not transpire! One undeniable consequence of the preprint is that it offers instant gratification to authors who wish to make a prior claim, to avoid competition and not be eclipsed by others. Scientific research is highly competitive in terms of funding, facilities, jobs, etc., it is little wonder that the preprint is an attraction for the bold-hearted (younger) researcher. As one positive author remarked – 'I made my mark in the sand' (clearly forgetting what the next tide would do).

The content of preprint papers is noteworthy. The format suits best physical sciences, bio-informatics, bio-engineering, computational and evolutionary biology (i.e. more databased research), but much less the life sciences with its high content of experimental (hands-on) research. Following on from this and for reasons of security and anonymity, preprints seem to be discouraged in the medical sciences.

To conclude, the preprint has found its place in the scientific world of journalism; it certainly has its drawbacks, but it does offer a means of quicker and more direct interaction of researchers with one another. A comparison often drawn helps us to understand its role and purpose. Imagine that you have been preparing a draft of your new findings before attending a conference. Your results can be presented in the form of poster, and accordingly you will meet others interested in your work who will discuss the findings with you on the spot. This is valuable feedback that will almost certainly help in reassessing, redoing and improving your contribution to science as and when it turns into a submission to a learned journal. A preprint can help you achieve a similar outcome to the poster, but your audience is not perhaps just a handful of conference delegates; it will be the world at large. One might be forgiven for thinking that a preprint has in fact to be seen as 'work in progress', but after all even a paper in a learned journal is 'work in progress' – scientific progress is open-ended.

12 Peer Review
The Crux of the Problem in Publishing Papers

12.1 Introduction

Peer reviewing deals with the problem of checking the scientific soundness of a piece of research being reported, that it is in its proper context, reads well, presents the data in an acceptable manner and concludes, after appropriate discussion, in support or otherwise of the hypothesis under examination. It also can be the stage at which plagiarism, duplication, falsification, fabrication and other unethical practices are most likely to be detected (see Section 12.2). It can be a tedious process, but it is the same for all authors, i.e. that their peers carefully assess whether a piece of research is going to be properly reported. In this sense, it is the most valuable element of quality control, so necessary in avoiding the scientific literature becoming swamped with second and third rate papers. The crux is that this is where many papers either succeed or fail – the latter are those that are scientifically and/or presentationally unsound. However, it is more likely that peer reviewers will receive reasonably sensible papers for review, as will be explained in the following subsections. Its value is not simply in deciding whether a manuscript is worthy of being published, but in making *constructive criticisms* regarding a paper's content and presentation, such that it can be improved by the authors, leading to a much better research article in due course. This is an invaluable job – a truly independent assessment of some piece of research by experts in the field. Much of what a person writes in a paper ought to be done with the idea firmly in mind that it will be read by experts. Their advice should be taken very seriously, although not all of it may be heeded. There is often a stage ahead of peer reviewing, referred to as *triage*; we must first consider what happens when a new article 'lands on the journal editor's desk' – or in today's terms, comes into the 'new submissions' folder on the editor's computer.

12.2   Triage

Not all papers are sent out for peer review, although this will depend on the policy of the journal. In those that are very busy and have limited staff to handle hundreds or thousands of submissions, it is better to weed out unsuitable and substandard papers as a first step. As a result, well over 90 per cent of papers may be rejected by journals such as *Nature* and *Science*, but substantially fewer will fail to proceed in journals where the impact factor is lower. Authors of papers rejected at triage will soon receive a decision, and thus they can move on to consider other journals without much delay. If your paper does not have something really important to tell the scientific world, it is very unlikely that you will have success with a high impact factor (IF) journal, and probably little success with journals of lower IF. There are some editors who, if not too inundated with papers, may wish to send almost every one of them out for review, i.e. almost all are given the benefit of the doubt, except papers that are clearly not in the subject area of the journal. On this account, and to save yourself embarrassment, resentment, frustration and ultimately time, think first about the level of your paper (its real substance and message), and the appropriateness of the journal to which you are submitting, so that it is written from the start for the most appropriate readership.

Triage leaves the editor with good papers, a number of papers to dismiss out of hand for a variety of reasons (including those mentioned earlier), and a batch in which the merits may not be so obvious. The last type can be sent to editorial board members for their opinion, who will advise the editor on the suitability or not of a paper, i.e. whether it can be allowed to proceed to peer review.

Triage is also a stage at which editors and their assistants can take stock of the papers being submitted for another reason. Software is available, operating at ever increasing levels of sophistication, that can track down duplication, multiple submissions, fraudulent and fabricated papers; indeed spoof papers coming in from unethical sources trying to invade the journals that have a bona fide existence (unacceptable 'predatory' behaviour) can now be tracked. This practice is becoming increasingly common, to which the only answer is more widespread use of surveillance and even better detection methods.

With regard to plagiarism (using/reproducing other people's material without permission), this is unethical and not tolerated in scientific publishing, as is reiterated here intentionally because it is so important. Software is now in use to deal with it (e.g. ithenticate.com by Cross Check), and is becoming ever more sophisticated.

12.3   Peer Review

We have reached the stage where papers that have not been rejected at triage will be sent out to experts for their critical comments. Peer review can be one of the most harrowing processes in the editing and publishing business, but it is one of utmost importance in maintaining the high quality of scientific articles, and also in helping to improve articles that are not quite ready or good enough for publication because of errors in methodology, poor interpretation, incorrect presentation of the data, missing controls and so on.

It is as well to remember that reviewers' time is precious and these busy people (yes, they are almost always the busy ones!) do not like being sent poor, weak and badly presented papers, especially those that they can quickly see as being scientifically suspect or frankly unsound. Papers that seem uninteresting also tend to be put aside and forgotten about, resulting in a long delay in reviewing time. It annoys reviewers to receive poor papers, often to such an extent that they will start to refuse reviewing for a journal that keeps sending them such papers. This is the unfortunate circumstance where a journal has the policy of sending out all submitted manuscripts for peer review irrespective of their basic merits, i.e. they do not bother to take papers through a triage. Good reviewers are too valuable for an editor to lose.

12.4   Choosing Reviewers

There are a number of ways in which reviewers can be selected, but it is generally not an easy job. The most obvious way for some of the bigger journals is that the editorial board is large, with experts covering most of the main areas in a discipline. They may even be paid a small amount for the work, although mostly it goes unpaid. Where a paper does not easily fall into the remit of one of the editorial board, the editor-in-chief will use the same procedures that most other journals use. There are three main ways of engaging reviewers:

(1) The editor maintains a list of reliable reviewers who have served him or her well in the past, from which the ones with the best knowledge of the subject matter will be chosen.
(2) The submission system automatically links up with papers that have recently been published in the same field by using keywords. Some of these will have come from reputable research outfits with acknowledged experts in the field.
(3) The authors may have indicated who would be suitable as independent reviewers, also noting those that would not be suitable for reasons such as a conflict of interest. (This matter was covered in Chapter 11 on Submission.) However, an editor must be astute where suggested reviewers may not be sufficiently independent or totally impartial. In this regard, nominated reviewers tend to be involved only when the other approaches seem to be getting nowhere, and their reports are carefully scrutinized for fairness by the editor, since there might usually be one report from a truly independent reviewer with which to compare it.

12.5 Contacting Reviewers

The submission software used by most publishing houses allows the editorial staff of a journal to access reviewers by email, inviting them to review a paper, sometimes allowing them to see an abstract, but usually in today's high-speed IT world allowing them to open up the whole document so that they can decide whether or not to agree. I mentioned that reviewers tend to be busy people, and while they are often the best, it is not easy to engage them.

Regarding this willingness of potential reviewers to read submitted papers, there are going to be many declining to review weaker papers, whereas very high quality papers with few corrections needed (those that might be seen in some ways as truly advancing the field of knowledge, and that are sometimes eagerly awaited by the top people in the subject area) will get priority from the experts, who might return reports within a few days. If this is done online, it has been possible to receive two favourable reports within a single day, but this is highly exceptional! Some journals have adopted the policy that reviewers should be identified so that authors know who produced the report. Others still do not allow reviewers to be identified. Often authors, particularly in very narrow topics progressing rapidly, will have a very good idea of who reviewed their paper because so few experts are likely to send in a reasonable assessment to the editor.

Papers that are not as outstanding are more of a problem – and this goes for probably the majority of submissions. More reviewers have to be approached to find at least two to respond positively, but the turnaround time is often weeks, perhaps months, despite reminders (about every 1–2 weeks, generated by the software). The excuse that is most commonly given for declining to review a paper is that (understandably) the person is too busy writing grant applications, travelling abroad, etc. Thus many papers are reviewed by people who are not going to be the highest flyers in the field. There is a moral issue in this regard; if these high flyers have such good publication records, their papers must have been peer reviewed by other experts in the field. In fairness we should all be prepared to help each other – the 'I'll scratch your back if you scratch mine' principle, but it does not work so well in peer reviewing.

The biggest delay in the passage of papers through to an editorial decision, therefore, is in getting hold of two (or more) reports from reviewers within a reasonable time. Conflicting reviews mean that editors have to turn to additional independent reviewers. This is also where the editorial board of some journals becomes much more useful. If one report has come in about a paper and a second seems to be unduly long in being submitted, the editor will turn to a colleague on the editorial board and ask for a report within a few days. By doing so, the length of time a manuscript is in the peer-review process can be kept within reason. In some of the biggest and most influential journals, there is a quite extensive editorial board covering all sub-disciplines of the field, and its members may well be under contract to the journal. In this case they receive some remuneration and are duty bound to peer review a certain allocation of papers per annum.

The problem of finding suitable reviewers

The biggest delay in the passage of papers through to an editorial decision is often in getting hold of two (or more) adequate reports from reviewers within a reasonable period of time. They must be experts in the field, who are often busy researchers.

Some journals have good review turnaround times, but some (notoriously, journals dealing with topics such as the history of science) can take many months and sometimes even a year. On this account, it is a good idea for authors to look carefully at the composition of the editorial board before making a final choice of the journal to which they might submit. A practice that seems to be declining is one of journals publishing the date of first submission and the date of acceptance of papers. When this information is given on the first page of papers in a journal, you can get some indication as to how long it took the paper to complete the editorial process. We also noted earlier that, post-submission, papers can be tracked online by authors, and they will often complain to editors when they see that reviewing is taking too long.

12.6   Reviewer Reporting

Peer reviewing is supposed to be impartial, confidential and thorough (not cursory). A great deal of trust is put in the peer reviewer who is in a very privileged position, and this, on occasion, can be abused, which may result in some very unhealthy and unethical practices that cannot be ignored in the editorial process (read more on this in Section 12.8).

Reports by reviewers have several sections, with some involving quick tickboxes and one-line responses, which are often less than helpful to the editor (except on the occasion where a paper turns out to be either dreadful or superb!), since they give too little specific information about the particular paper under review. Good reporting will start with a *short note* indicating the nature of the article being reviewed and a note as to whether the authors have indeed dealt with the subject matter and hypothesis as they intended.

It will move on to *general comments*. The reviewer is charged with the duty of indicating whether the paper is scientifically sound, that the hypothesis being tested is reasonable, that the design of the work/experiments is appropriate and that the data accurately record what was found. Statistical analysis will have to be checked to see whether the data have been handled properly. If the reviewer is not competent to advise on the appropriateness of the statistical tests and their accuracy, this should be intimated to the editor, who may then seek extra advice from a qualified expert. The discussion of the results has to be lucid, the relationship of the findings to other similar reports in the same field of research must be germane and unbiased, and the speculation has to be within reasonable limits. The conclusions drawn need to be checked for their veracity. This section of the report is the meat of it, which is followed by the specific comments.

Specific comments can provide several types of critique. One will be that of indicating where in the general comments issues may have been found that do not seem to ring true, and these are more carefully identified and discussed. It is possible that, after saying in the general comments one part of the evidence for the hypothesis did not seem to be true, the exact nature of the problem can now be dealt with in greater detail (e.g. that the data in 'figure 3' do not seem to support the hypothesis because the authors did not measure the appropriate parameters). The specific comments can come down to a paragraph-by-paragraph, line-by-line account of errors or preferred options that the reviewer wishes to point out. There are some reviewers who go to great lengths in this regard, which is very good news for the editor where a paper might go on to be acceptable, but would otherwise require very extensive editing. The authors will in due course receive this information and are required to produce a much improved revision if a paper is *provisionally accepted*. Thus peer review raises the standard through critical appraisal of work done by others. There are two forms of criticism, *constructive* and *destructive*. If a paper is badly presented and unsound in its content, it is not destructive criticism to say so; it is simply the truth. Destructive criticism is where much of a paper is inherently wrong, misleading, inaccurate, and so on, and the reviewer takes the authors to task step by step without giving any real help. Authors would be demoralized if things are simply said to be wrong, wrong, wrong. If a paper is bad to the core, this will be given in the general comment and only a few specific points might be needed to indicate places where things simply do not add up. A weak paper with some merit in it, on the other hand, can be greatly improved by constructive criticism. Authors need this type of feedback on their work, and good commentaries on papers should be gracefully and gratefully received.

The use of tickboxes is not necessarily a blessing, but can be one of the problems with reviewer reports, making some of the categorizations of a paper too tight; however, most tickbox sections do allow for 'additional comments'. These boxes help the busy editor because he or she can scan reports quickly and get a good idea whether a paper is weighted more on the acceptance than the rejection side. The **Electronic Reviewer Report** gives you an idea of a typical reviewer report form which, on completion, will be submitted online back to the editor.

A brief remark on constructive and destructive criticism

A weak paper can be improved when constructive criticism is given, which will encourage authors to revise their manuscript. But authors are quickly demoralized if things are simply said to be wrong, wrong and wrong again through destructive criticism coming from reviewers.

12.7   A Typical Reviewer Report Form

Figure 12.1 on the next page is a report form on a paper dealing with bladder cancer, which gives some idea of the way in which a reviewer needs to respond to the editor of the journal by filling in the online template with his or her remarks that will go towards deciding the fate (along with another report from a second reviewer) of the paper.

12.8   Abuse of Privilege in Peer Reviewing

When an expert reviews another person's paper, he or she sees information that must remain *confidential* until it is published. Discussion of the content of this paper with the reviewer's colleagues and others where it

```
Prof xxx has returned a report on this manuscript. Please see the comments below.
MS ID  : ….1123
Title  : A proteomic analysis of metastatic potential heterogeneity of muscle invasive bladder
cancer and biomarker discovery from urine
Journal: Cancer Cell International
Authors:…..

You can view or change the status of any reviewer for this manuscript via:
http://www.cancerci.com/editor/manuscript/peerreviewers/view.do?manuscriptId=xxxxxxx0

Reviewer's report-
General remarks: The authors examined protein expression profiles of urothelial carcinoma of
bladder (UCB), an important malignancy in terms of frequency. The authors have identified
molecular backgrounds of malignant features of UCB, and tumor proteomes with different metastatic
risks. One of unique points of this study was that the authors focused on tumor stroma proteome
using laser microdissection and iTRAQ. Interesting proteins were identified, and classified
according to their possible functions.
The approach is substantially interesting. The importance of tumor stroma in poor prognosis is
well recognized but has not been extensively examined by proteomics. This study employed laser
microdissection to overcome tissue heterogeneity. The proteomics modality, iTRAQ, is an
established protocol, and the results appear sound.

Specific comments: There are several points for Minor Essential Revisions.
First, the purpose is for a basically deeper understanding of molecular backgrounds of metastasis
risk, which may compliment the drawbacks of TNM classification. But the outcome is mostly a
classification of identified proteins, and needs to contribute more to our understanding of
metastasis in USB. Some in vitro experiments about the proteins, which were not reported in
metastasis especially in USB need to be done. Alternatively,………
Second, it is not clear whether the authors examined samples of the individual cases or pooled
samples. The authors should clarify this point and discuss possible problems. Cases in the same
category may have different molecular backgrounds…. I wish the authors could identify proteins
which stratify the patients according to the protein expression pattern, even in the same risk
group….
Third the number of proteins was 1049, and I am wondering what percentage of proteome was covered
in this study. Obviously, 1049 is not so large… The authors should have discussed the technical
limitations…
Fourth the idea of linking tumor and urine proteomes seemed to be very attractive. However, there
are a lot of proteins from different tissues/organs in urine, and it may not be so simple to link
these two proteomes. The authors found some reasonable correlations, but it may be just
coincidence, because they observed many proteins. In this situation, false positive correlation
can easily happen.

Level of interest: An article of importance in its field
Quality of written English: Acceptable
Statistical review: No, the manuscript does not need to be seen by a statistician.

***********
Confidential comments to editors: The aim and approach in this paper is quite interesting, and
worthy of publication. However, the results were not entirely sound. The authors classified the
identified proteins according to the GO classification, and discussed about the results of
classification and individual proteins. It is difficult to conclude that this paper contributes
significantly to our further understanding of molecular backgrounds of metastasis of UCB. On the
other hand, the protein list in this study will be beneficial for the readers and researchers who
are interested in the technical aspects of this approach and the proteome of this malignancy. The
urinary proteomics part of this study should be improved by revision. I suggest acceptance of this
paper after minor but essential revisions.  In particular, the authors should discuss the
limitation and possibility of their research.
What next?: Accept after minor essential revisions
Declaration of competing interests: Nothing to declare.
```

Figure 12.1 Electronic reviewer report.

shows something particularly interesting or novel should not occur, but inevitably to some extent it does. Taking this to extreme, the information in such a paper might be so useful to others in the same field of research that they use it to their advantage before it becomes public knowledge (i.e. is published). We will discuss this further in Chapter 16, but it is worth noting here that the passage of a paper might be intentionally delayed so that a scurrilous peer reviewer can take advantage of its content, and make use of it to his or her own benefit. It might be trivial, but in the worst cases, it comes down to *stealing* ideas and information, and using them for unfair 'profit', e.g. someone could write a (successful) grant application on a new hypothesis proposed by the unsuspecting author whose paper fell into the hands of an unscrupulous peer reviewer. Although this is a warning that these things can happen, fortunately peer review is largely an ethical and constructive exercise.

13 The Last Stages of the Editorial Process
Decisions, Revisions and Final Editing

13.1 Introduction

Earlier chapters, especially the previous one, have dealt with several issues on the editing side. Without being repetitive, the first steps have been dealt with in Chapters 11 and 12 – both the submission of papers online into the system used by the chosen journal/publisher, and the business of first screening (triage) and peer reviewing. What follows thereafter is largely in the hands of the editor.

13.2 Editorial Decision

When reports have come in from reviewers, the editor and editorial board will carefully scrutinize and compare them to determine the probable fate of a paper. When two reviewers send in differing reports, it can easily be spotted whether one has done a good job and perhaps the other has given a poor critique, i.e. they are at odds with one another. In this circumstance, there is only one course of action, which is to get a third (or more) independent review(s). Editorial board members can help, since they can be asked to arbitrate and act as further reviewers who are expected to return their verdict quickly – otherwise a difficult paper may be locked in the reviewing stage for longer than necessary. As mentioned in Chapter 12, modern submission systems allow authors to track the progress of their articles, and many complain when their papers spend too long in review.

When the reports allow the editor/editorial board to make a decision, there are several options available for papers that have been *positively* assessed. The best outcome by far is for a paper to be seen as of high merit, well written and can be accepted 'as is'. Unfortunately these are relatively few and far between. The majority fall into three groups:

(1) Acceptable – usually subject to appropriate minor corrections – seldom is a paper perfect!

(2) Provisionally acceptable – subject to major revision, for which about 2–3 weeks might be considered an adequate time. The revised MS as a resubmission will be carefully checked by the editor, or when necessary, by one of the original reviewers.

(3) Unacceptable – but occasionally with the option to submit a new and greatly improved manuscript on the same topic, i.e. a nod of approval from the editor suggesting it may in future have a better chance of acceptance.

This last category might be a paper that is of considerable interest in what it purports to tell, but is in need of better controls or further experiments that can move it from class (3) to class (1). The editor is suggesting that the authors ought to have more time to complete the revision properly. In these circumstances, the editor has a further two options: (i) allow considerable time (not weeks, but perhaps 1–2 months) for the authors to complete the work and revise the manuscript, or (ii) suggest that the manuscript is withdrawn, but that a *new* submission would be welcome when the work has been properly completed. In both cases, especially the latter, the revised/renewed manuscript would need to be seen by one, if not both, of the reviewers again before the editor can decide whether or not the paper is fully acceptable. In these circumstances, triage could have helped since it can weed out those papers that are unlikely to be accepted without extensive revision. Clearly editors are not in a position in some cases to make such a judgement, and therefore the author gets the benefit of the doubt in the paper going out to review.

On the negative side, there are also couple of options:

(1) The paper is given an outright rejection, and there is usually little or no redress in these circumstances (but see Section 13.3 regarding *rebuttals*).

(2) The editor tells the authors that the paper is insubstantial and has to be rejected because of the unfavourable comments of the referees. However, on occasion there can be sufficient merit or new material in a submission to suggest to the authors that a short communication or preliminary report on its basic finding be prepared, which might be worth reconsidering by the editor.

13.3   Dialogue with Authors

Having made the decision, editors now have to get back to the authors regarding the fate of the paper. For the papers that came through positively, most authors will find the reports of the reviewers sent back to them very useful. They see the versions of the reports written by the reviewers specifically for them, while there will be other confidential comments that only the editor will see.

It is at this stage that authors can now comply (or not) with the requests for revision. In the vast majority of cases, authors do revise their articles appropriately and resubmit them as requested. Others, given the option of a resubmission where extensive revisions are requested, may need more time to complete extra experiments, as mentioned earlier. However, there are occasionally authors who decide that they will take the advice but then seek to publish their improved papers elsewhere by submitting to a different journal. This is unfortunate for the journal editor because he or she will not necessarily know what the authors are doing with their paper, because there is usually no response. The editor will have lost a potential publication after all the hard work invested in it. Unfortunate as this might be, it is all part of the publishing process and has to be tolerated.

Regarding rebuttals: There are some cases that present additional problems for editors, especially where authors receiving the reviewers' reports find the criticisms unwarranted and sometimes frankly untenable. Every author has the right to present a rebuttal to an editor regarding a paper that has been unfairly criticized, e.g. often being asked for major revisions that are inappropriate or quite untenable. The reviewers might have been too unfamiliar with the field and therefore ought not perhaps to have been invited in the first instance. By entering into dialogue directly with the editor (i.e. it is better not done through the submission system online), authors have this option of redress, provided they can cogently argue their case and perhaps corroborate their views through other independent experts. A complete reappraisal may be needed, a new set of reviewers appointed, and possibly several editorial board members asked for their advice. The editor will have the final say when all the reports have been collated.

There can be a genuine misunderstanding between what has been presented in an article and how the reviewer has interpreted it. Some of the most outstanding papers that have made major breakthroughs in science have suffered at the hands of peer reviewers. If something very new has turned up, even the pundits may not believe the data and the explanation; they may have too little insight to see something that is radically new, different and important. However, that is usually an exceptional scenario. On average, rebuttals relating to more average articles are very infrequent.

13.4   Revising Papers

Authors should be thorough in their revisions. Some authors carefully itemize each response made to individual comments, ensuring that the appropriate adjustments can be seen to have been done. Other authors highlight text and places where revisions have been carried out. By and large, most authors are assiduous in completing their revision, but there are a few who make some of the amendments and ignore many others. Authors should therefore not assume that a revised manuscript will be fully accepted unless they are seen to have been thorough. Doing a poor job simply means that the paper may be passed back and forth between the editor, the reviewers and authors several times, which should be avoided as it simply wastes a lot of time for everyone. An editor can quickly see if an author has indeed made a good job at revising a paper. I have had one author claim to have thoroughly revised a paper and yet returned a completely unaltered manuscript (and not by mistake!). When an editor is finally satisfied, the paper is accepted, and its latest version will then be prepared for publication.

Small corrections: Modern online publishing means that the submission systems allow authors, editors and publishers to deal with the manuscripts right up to the time of publication. Editors will always work with the latest version in the system. Substantial changes to the substance of a paper, other than suggested corrections, cannot be made. For example, you cannot upload an updated version that contains new experimental work that has unfolded in the interim, since this would require that it has to go back through the review process again. Small changes (errors such as incorrect orientation of the axes of a graph, a wrong symbol that needs to be replaced, etc.) can be made right up to the time when the various formatted files of the paper are being converted to the final HTML or PDF format.

13.5   The Editor's Job in Creating the Final Version of a Paper

It is easy for a well revised paper to go ahead to the publishing step if it has been prepared by the authors in strict accordance with the Guidelines set out by the journal. Each journal will have its own rules and the job of the editor is greatly simplified if authors have fully complied. Each paper will be carefully checked for completeness, soundness and clarity of presentation, and the editor will make the final version ready for the publisher.

There is only one good way in which to show how a final edit of a paper can greatly improve its final form. A file is available (see below for the link) which shows the main text of a paper that has gone through the most thorough review (an extreme example), but it makes the point that the editor is there to ensure that the final version of an accepted paper has both excellent presentation and sound science. The file can be found at www.biomedes.biz (under Menu, choose Specimen Paper).

13.6   Preparing for Publication

By the time a paper has reached this stage of completion, the editorial job is to ensure the manuscript reads well and is in line with what is required by the publisher. The publisher then takes the manuscript through the final stages ready for printing (hardcopy); however, journals are nowadays almost exclusively published online. The closer the manuscript coming from the editor is to the exact requirements of the publisher, the easier and quicker its passage will be. These final steps concern the very exacting job of setting the paper in the right order, checking in detail every bit of the formatting (e.g. the typescript sent by the editor will not be in double-column format, some fonts will be changed, especially where footnotes are being used, the legends to figures need to be correctly attached and the positioning of all figures, tables, images, etc. made final). Also, any further details that are appropriate will be added, such as generating a digital object identifier (DOI; see Section 8.8 and Chapter 14). All these give the paper the distinctive character of the journal in which it is to be published, and will be described in the next chapter. The publishing editor will see to these steps, and the final proof will be returned to the author to check and amend as and where necessary before the proof goes to print.

14 From Acceptance to Publication

14.1 Introduction

An accepted paper is checked again for originality, because of the possibility that it or a very similar one has already been published. This also eliminates any instance in which the paper has also been accepted and published by another journal, which is ethically unacceptable and will lead to the paper being rejected. The authors and their institution(s) will be informed of this malpractice (see Chapter 16) and serious cases can be sent on to the Committee on Publication Ethics (COPE), which makes it possible to openly discuss discretions online. The Committee on Publication Ethics has guidelines set out as flow charts on how to deal with unethical practices.

In times gone by, printing a paper was a tedious process of compositing – changing the written article into a printed page that had to be set together with lead pieces with the font on them (back to front), inked and pressed on paper (the 'printing press'). This was an exceedingly time-consuming and exacting task which thankfully has gone. In those days, it was best practice to keep papers as short as possible to get them through this arduous stage as quickly as possible. Today, the author has the main responsibility for composing an electronic version of the paper; the more precisely it follows the journal's instruction, the easier and quicker it should get through to publication. But this does not excuse excessively long papers, even though it is now possible to publish online *in extenso*, an almost infinite number of papers. Probably one of the most annoying things an editor meets is when a submitted paper turns out to be very good and publishable, but has probably been rejected by another journal and the authors have not bothered to change its presentation to that required by the second journal to which it has been submitted.

As indicated earlier in this book, the procedures from the editor's office and then through the publisher's office involve considerable work. Thus short papers are more welcome than tediously long ones; it also generally means that their authors have probably been more succinct in their presentation, making them clearer and easier to read. Authors often disregard the reader; papers should be written for the sake of the reader rather than simply to gratify the author. Readers are generally busy people who do not have the time to wade through long and often boring papers.

14.2   Procedures in the Publishing House

14.2.1   Format Changes

Although manageable, some authors have the habit of submitting papers, or parts of papers, that are relatively inaccessible and therefore increasingly more difficult to edit. Authors should check that all of their submission can be edited.

An accepted manuscript is needed in both Word and PDF form; the editor will already have made corrections in creating the final version, which is as far as it can go before handing over to the publisher. It is meticulously checked again for errors in spelling, punctuation, etc. when it reaches the publishing editor. Consistency in the language is also checked; for example, a paper has to be written in UK English or US English, not in a mixture of the two.

The next exacting task for the publishing staff is now to get a paper into the final *format*, which means that all fonts, spacings, headings and sub-headings, footnotes, reference listings, etc., must be correct. Although authors and editors present papers in a particular way, this is not precisely how the finished published product will appear. Each publisher will have its own preferred style of presentation. Here are some examples of the differences between the accepted paper and its final published form:

(1) manuscripts will have been written with only one column per page, but very many journals use a two-column format;

(2) most journals justify the right as well as the left margin;

(3) the corresponding author's details and other items may often be found in a footnote at the bottom of the first column on the first page;

(4) subheadings might be numbered and in a different font;

(5) the references are checked so that they are in the correct style;

(6) line spacing (often 1.5 or 2.0 in a manuscript) is now reduced to 1.0.

There are many other very small details that must be dealt with, some of which are mentioned in the following paragraphs. Most authors put the Materials and Methods section immediately after the Introduction; however, in a number of journals the order of the sections can be different, and you may find it after the Discussion section, i.e. at the end of the text. It might also be in a smaller font.

Here are some finer details that need to be considered in producing the final form of a manuscript. In a lot of papers, hyphens will be missing. Double spacing may have been used between words or after punctuation, i.e. the space beween words may be more than one 'el' (the space introduced by a single touch of the space bar on the computer keyboard). This is a common mistake that ought to be picked up by using the MS Office Word program during the writing of a paper.

The fonts used may be erratic throughout a paper; for example, the text font might change in type and size from one paragraph of the text to another. The publisher will ensure consistency throughout the final publication, but much of this work is eliminated if authors take due care and attention in writing the paper.

Subscripts, superscripts, unusual letters (e.g. ñ, used in Spanish, would not be appropriate in an English text, unless it is in a name) and similar minor details need to be corrected. During changes from one format/ program to another, subscripts may have reverted to normal text, e.g. C_n may have become CN or OC may occur if using superscript 'o' (^{o}) rather than the degree symbol ($^{\circ}$).

14.2.2   Figures, Tables, Schemas and Others

These items have to be inserted into the right places and the legends appropriately attached. Some authors will have indicated where these items should go, but were asked to submit them as separate files. However, Figure Legends are generally listed together after the Reference list and do *not* accompany the figures. A figure (in an approved format, e.g. JPEG) will finally be positioned close to its textual reference, but in such a way that it can be suitably accommodated in the one or two column set-up. It might be changed in size to make it fit on one column, but sometimes they have to straddle two columns. The ratio of the height and width of a figure (already discussed in Chapter 9) ought not to be altered by the publisher, but the overall size can often be quite radically altered. This has caused much concern, as previously discussed, because one can lose much definition in, for example, an electron micrograph. Reducing the size of some complicated graphs might lead to difficulty in distinguishing one curve from another.

One of the biggest complaints has been with fluorescence images. In many cases a series have to be compared, and some (indeed most) publishers started to reduce these to thumbnail size, making the detail within them almost invisible. While colour printing of hardcopy used to be expensive (and remains so), the cost online is negligible, and small figures can be enlarged by zooming in, with the only difficulty arising where the original resolution has been reduced or degraded (fewer pixels).

Many figures are not cropped by authors; this means that a lot of white space is included around them. This should be dealt with at a much earlier stage by the author rather than leaving the publishing editor to deal with the problem.

The relevant figure legend is then appended beneath a figure. Occasionally figures and/or their legends can get muddled up, which may well require to be sorted out with the author, preferably before the final proofing stages.

Tables are best viewed when on a single page, not continuing over to the next page. Sometimes this problem can be easily resolved, e.g. by changing the orientation to landscape from portrait. When this does not work for exceptionally long or wide tables, it is often possible to break them up into sections dealing with specific issues within the same context, thereby creating two, three or more separate sections that are more easily assimilated. The same goes for overcrowded figures, i.e. those with far too many curves on them to separate easily from one another by the reader, where clarity is achieved by making several smaller figures in which the control and only two or three of the curves are used (refer back to Chapter 9 for more information). The issue here is that if things are left to the publisher to edit/manipulate, this may lead to some dissatisfaction that could take some time to sort out at the galley (uncorrected) proof stage and therefore is better being dealt with at the writing stage, or by being explicit in what you require by discussing matters with the editor and/or publisher, leaving specific instructions with them.

Complex equations are centred usually in a column, but sometimes disappear to be replaced with two question marks or empty boxes in the proofs. The authors will have to deal with these problems by converting them to a form that can be seen within the text. Once introduced into text boxes, some equations drawn up by a mathematical program cannot easily be manipulated, and clearly unusual symbols can be present that do not translate from Word or some other program as the paper goes into its published format.

When everything has been done by the publishing editor, and the component parts of the paper are integrated into one file, the paper has now moved on to the *proofing* stage. It is now ready to be *copyedited* by the publishers either in house or through a subcontracting firm that takes on this role. There is a very good reason for the latter, which arises because overfamiliarity with such a thing as a learned document can lead to errors being overlooked time and again – we might be said to become 'word-blind' to them. This is why an independent reader who has never seen the document before can often spot errors or omissions that were missed by the publishing editor.

14.3   On Completion of the Proof

The proof will be usually in an editable format, such as Word or an online editable system. This allows the editing of text and comments about the figures or other paper attachments. The online system also allows copy-editing questions to be answered by the author. Some publishers only send out PDF files as proof copies, and PDF file annotating software is necessary to alter these files (sometimes supplied by the publisher). These files are convenient to pass back and forth between the editor and author to eventually move a paper from a hidden site only accessible to these parties before it goes online or for final printing. A URL is created for this purpose and both editors and authors can proofread it. When the final corrections have been done (see the following note), the paper will be given a volume number and either page numbers or a paper number for that volume (paper 1, 2, 3 and so on, as they get published in the volume). The URL is added in the final version and a DOI (digital object identifier), which is a number that can be clicked on to allow a reader to download the paper after it has been put into its final form and uploaded to the main website of the journal. It now becomes available to everyone worldwide, either free – *the author pays* on acceptance – or after paying a subscription – the *reader-pays* system – more often than not through institutional payments to the publishers.

Note: at this stage, it is usually only small errors or additions that are allowed unless some vital information is needed to make sense of some part of the paper. Extensive revision is seldom allowed; an author reaching the conclusion that this is absolutely necessary might be told that the paper will have to be resubmitted or even withdrawn so that a new submission can be received.

As an article goes online, an automated email informs both the authors and editors that it is now 'live'. The paths of most manuscripts through the publishing process are set out in Appendix 14.1 as a flow chart. The smoother the passage of a manuscript through the stages, the sooner it will be published. Appendix 14.2 details the types of files involved.

14.4   In Summary

- Publishing editor receives paper and checks it over.
- Editable files are sent to the copyeditor for formatting.
- Proof finished by copyeditor is sent online as a PDF to author for checking.
- Proofed copy sent back to publishers.
- Additional comments added and any manuscript reformatting completed.
- Online manuscript, unavailable to general public, can still receive very minor changes (even by author).
- Online manuscript agreed, completed and assigned a DOI can be pre-published online at this stage.
- Manuscript assigned journal volume and page numbers and fully published online and/or in paper copy for distribution.

Appendix 14.1 Flow Chart

The Final Steps in Publication

Author sends in the finally corrected article after proofreading

↓

Publishing editor sends the edited version on file to copyeditor

↓

Publisher sends returned approved file to authors (the galley proof)

↓

Authors check proof with their final corrections

↓

Publishing editor checks corrections

↓

Proof sent on for copyediting which ensures

That all the text is in the correct format of the journal

↓

Final manuscript put online on publisher's server and assigned a DOI, and volume, issue and page numbers

↓

Final manuscript published online

Appendix 14.2 Description of the Types of Files Commonly Used

Portable Document Format (PDF) The native file format for Adobe Systems' Acrobat. PDF is the file format for representing documents in a manner that is independent of the original application software, hardware and operating system used to create those documents. A PDF file can describe documents containing any combination of text, graphics and images in a device-independent and resolution-independent format. These documents can be one page or thousands of pages, very simple or extremely complex with a rich use of fonts, graphics, colour and images. Although conversion of PDF to editable forms is now becoming easier, very minor changes can be indicated by using the 'sticky note' option on the menu bar of Adobe.

Rich Text Format (RTF) This is an interchange format from Microsoft for exchange of documents between Word and other document preparation systems. However, a well presented paper in Word may become quite differently displayed in RTF because the formatting is different (e.g. right justification is usually lost).

Device independent file format (DVI) A DVI file containing a description of a formatted document is the usual output of TeX. A number of utilities exist to view and print DVI files on various systems and devices. Mainly used from Macintosh platforms.

Encapsulated PostScript (EPS) An extension of the PostScript graphics file format developed by Adobe Systems. EPS is used for PostScript graphics files that are to be incorporated into other documents. On some computers, EPS files include a low resolution version of the PostScript image. On the Macintosh this is in PICT format, while on the IBM PC it is in TIFF or Microsoft Windows metafile format.

Tagged Image File Format (TIFF) A file format used for still-image bitmaps, stored in tagged fields. Application programs can use the tags to accept or ignore fields, depending on their capabilities. While TIFF was designed to be extensible, it lacked a core of useful functionality, so that most useful functions (e.g. lossless 24-bit colour) require non-standard, often redundant, extensions. The incompatibility of extensions has led some to expand 'TIFF' as 'Thousands of Incompatible File Formats'.

Joint Photographic Experts Group (JPEG) This is the original name of the committee that designed the standard image compression algorithm. JPEG is designed for compressing full-colour or grey-scale digital images of 'natural' real-world scenes. It does not work so well on non-realistic images, such as cartoons or line drawings. JPEG does not handle compression of black-and-white (1 bit-per-pixel) images or moving pictures. Standards for compressing those types of images are being worked on by other committees, called **JBIG** and **MPEG**.

Bitmap (BMP) Microsoft Windows can provide the bitmap format. This is the only graphics format where compression actually enlarges the file. The format is nonetheless widely used.

15   Copyright

15.1   National and International Coverage of Copyright

Copyright is recognized worldwide, and international conventions guarantee that your work will be protected to at least a minimum level in most countries. *The Berne Convention* is the gold standard. Search its website for more information, but note that the World Intellectual Property Organization (WIPO) is preparing to update it (see the box and discussion that follows). The Berne Convention was only fully adopted after 1988, mostly because the USA had difficulties with certain demands from the Library of Congress. There are frequent changes in copyright guidelines and legislation, and several recent reports have led to new discussions (see Section 15.6).

General information on copyright

Copyright law is complex, but it is much the same worldwide. It applies not only to science and medicine but to:

- theatre;
- live music performances;
- recordings of any type;
- the world's literature;
- technical manuals;
- new inventions and drugs; and
- most other publishable material.

For artists, public speakers and others, copyright is normally arranged by agents.

In the UK, copyright operates under the *Copyright, Designs and Patents Act, 1988.* When you agree to transfer copyright to a publisher, your work is protected in all forms and media throughout the world (this comes under Section 80 of the Act).

Copyright lasts for 20–25 years for most purposes, including patent protection, but it has been extended to 75 years in some cases. The universal symbol for copyright is ©.

To paraphrase the basic tenet of the law: Copyright becomes an issue when a work is created, is seen as original and has arisen by 'labour, skill or judgement'. 'Interpretation is related to the independent creation rather than the idea behind the creation.'

The notion of an idea being protected by copyright needs more detailed discussion, especially in the realms of science and medicine; see below.

This translates into the International agreement referred to as the Berne Convention for the Protection of Literary and Artistic Works. It was first adopted in 1886 as an agreement to honour the rights of all authors who are nationals of countries that are party to the convention. The current version of the convention is the Paris Act of 1971. The convention is administered by the World Intellectual Property Organization (WIPO).

(Source: The Berne Convention for the Protection of Literary and Artistic Works (Paris Text 1971) – Article 7)

15.2   Science and Medicine: Author, Employer and Publisher

In Chapter 11 we saw that most systems for submitting a scientific paper have a section concerning copyright. By signing the copyright agreement you either:

- release your work so that the publisher will hold the copyright, or
- retain copyright yourself.

Most, but not all, journals/publishers insist on the former. It is usually easier for the publisher to take legal or other action following copyright infringement than for an author to do so. However, if you are an author of many scientific papers, you might hold copyright to some while your publisher holds copyright to the others. You need to be aware of the difference.

Usually, numerous conditions are associated with the Act and the conventions. When you assign your paper it will not progress into the public domain unless the publisher holds the copyright. Therefore, delay in transferring copyright will delay publication of your work. These conditions should be considered not only by the corresponding author of a paper, but preferably by all its authors. (All authors' names should be included in the submission form, or at least it should be stated that all the co-authors agreed to assign responsibility to the corresponding author.)

When a publisher holds copyright, it is an offence to publish the same paper or parts of it anywhere else unless permission has been sought and granted. It is most important that you *never use other people's work without first getting permission*; it is no good seeking permission afterwards. All the bigger publishers have permissions departments who deal with such matters on a daily basis (see Section 15.3). (We will deal with *plagiarism* – i.e. copying other people's work as if it were your own, without seeking permission to quote it – under ethics in Chapter 16.)

An author who is employed by the company or institution in which the work was done is normally obliged by contract to relinquish copyright to the employer. This means that copyright cannot be transferred directly to a publisher, so a licence is issued that allows the publisher to proceed.

15.3   Permission

Only the copyright owner has the right to authorize reproductions and adaptations of an author's work. The copyright owner is usually the original creator, or perhaps the next of kin, unless (as for most scientific papers) copyright has been transferred to, for example, a publisher. If you wish to make and publish a 'derivative work' – something like the original – you need permission from the owner, unless the original work has a licence that explicitly allows for derivative works, in which case specific rules and conditions must be met. Educational establishments can often be allowed to use your work, but here the rules depend on national legislation and differ from country to country. An awkward problem in this context is: how far can one paraphrase another person's work and get away with it?

Permission is usually obtained from the publisher of the work. If the publisher cannot give permission directly, they will know whom you should contact (in order to use the work in the first place, they must have been granted permission by the author, the company, the trader, etc.). There could be a private agreement, which is best put in writing to preclude future arguments or challenges. For website material, contact the webmaster, who may either give permission directly or refer you to the owner.

To speed up the process, you should:

- state how the work will be used, how it might be adapted or changed without being misleading;
- fully describe what you wish to use. This includes the name of the author and title of the work, the ISBN number (or ISSN for periodicals) if necessary, and an exact description of the content you wish to include (title, version, illustrations/images/diagrams, chapter/section/page numbers, start and end points of the extract required, etc.);
- state how much material might be copied and how it will appear in its new guise;
- ensure that the work is properly attributed. Thus the source of your material, the name of the owner and the exact reference must be included. In making the request you will need to give your full contact details (or your agent's, where applicable).

15.4   How the Passage of Time Affects Copyright Restrictions

In science, the common good is all-important. Most papers over 25 years old are seldom challenged over copyright unless they are seminal publications. Nevertheless they should not be quoted *in extenso*. Science proceeds by consensus, so it would be inappropriate to challenge anyone moving the field forward a generation or so on. Most of the scientific literature moves quickly out of copyright into the public domain, but you still need to acknowledge relevant major findings or innovative methodology introduced decades ago. Work that has entered the public domain becomes available to all; the authors can no longer stop others using it or claim copyright on it; nevertheless, courtesy and respect for our forebears are expected. Whatever the circumstances, it is essential to cite the source of the copied material (a full reference, the patent number, etc.).

15.5   Truly Innovative Acts

When you transfer copyright to a publisher, a patentable invention or procedure specified in the publication does not become the new owner's property. Any royalties from the invention will go to you and not to the publisher. However, you will first need to protect the new invention by applying for a patent before the article is published, otherwise the patent is void.

However, remember that an idea alone cannot be protected. There must be something tangible to protect. Stealing other people's ideas, particularly in science, is unethical. The original reference must be cited. Communication of new ideas is vital for science and the authors of all ideas – even wrong ones (the large majority!) – must be respected. New ideas that prove correct and important must certainly be acknowledged; this is the most important of the few tangible rewards in science.

Stealing someone else's ideas when they have been submitted for publication but are not yet published is much less easy to control. There are several ways this can come about. These issues just mentioned will also be mentioned under Ethics in the next chapter.

15.6   Authors' Licensing and Collecting Society

Income from publications that have to be purchased (rare for research papers, but usual for books) goes to the publishing house. The authors receive a negotiated royalty (often 8–10 per cent of net profit). Such publications can be read by people using libraries; they have not bought any right to purchase them. The Authors' Licensing and Collecting Society (ALCS) tries to redress this problem by collecting data on the number of times a publication is accessed by 'the general population'. This enables them to divide a pool of money among many authors, up to a maximum of several thousand GB pounds per year (see www.alcs.co.uk). This allows authors to at least get a token recompense from readers accessing their publications (sadly this does not happen for the vast majority of scientists, who ought to be satisfied simply by the fact that someone does read their papers!). ALCS also provides news about changes in copyright laws, which are under review and revision in the UK, with subsequent alignment throughout Europe being required:

> The current UK review of copyright is not taking place in isolation. . . .the European Commission published plans for its own review of copyright, setting out a two-year programme to look at a number of areas including the exceptions to copyright that Member States may include in their national legislation.
>
> (Source: ALCS website)

Global legislation must also be updated. This is necessary today since any publication can easily be disseminated throughout the world.

> . . . at a global level the World Intellectual Property Organization (WIPO) Copyright Standing Committee has been considering the question of copyright exceptions for some time and has recently included the issue of educational use to its agenda. There is clearly a growing impetus in international copyright discussions to explore ways to expand permitted access to authors' works. Throughout 2013, ALCS will continue to monitor and report on these developments, intervening and lobbying wherever necessary to protect the interests of writers.
>
> (Source: ALCS website)

16 Ethics and Scientific Integrity

16.1 Introduction

Scientific writing and publication requires that the whole practice is ethical and as transparent as possible. If we are seeking to gain knowledge that adds by consensus to our greater understanding of life and the Universe, then we cannot afford to have unreliable and frankly contrived data, especially when it is deliberately and not unintentionally reported. Such unacceptable behaviour and transgression of good practice might set back work on a hypothesis by years, certainly when it is not quickly evident that malpractice has been involved. It follows that scientific endeavour, including the writing and publishing of our work, has to be scrupulously accurate, clean and honest.

But scientists and doctors, like other professionals, are human. Life's pressures often put people under strain, resulting in some bad practices that must be recognized and eliminated. It is the job for all of us to do this if we are to maintain a high gold standard. In this regard the editing and peer reviewing processes provide quality control at a level that is probably stricter than in almost any other branch of publishing worldwide. This is being made easier every day by the implementation of new software that is capable of detecting fraudulent practices.

However, indiscretions do occur that get through the system, although general science, compared to many other fields of knowledge, is relatively free of them. Authors starting out to publish their work need to be well aware of unethical practices, and have to learn how to avoid engaging in any if they are to proceed in the most ethical way. On this account the best advice is to look very carefully at the 'Instructions to Authors' for every journal to which you may be submitting. As mentioned on several previous occasions, there will be a box in the submission system that specifically asks for your confirmation that your submission is not being sent else-where at the same time. If you do not comply, the publishers will not be interested in proceeding with manuscripts you might wish to submit.

Malpractices include: non-disclosure of a commercial interest in the contents of a manuscript; placing libellous remarks in a paper; disclosing someone's name in, for example, a clinical case-study (loss of anonymity); not disclosing a conflict of interest; and many others. But we need to examine the main ones more closely in the following sections.

16.2   Highly Unethical Practices

16.2.1   Duplication

A paper that deals with new findings or ideas that can advance our field of knowledge is expected to be novel and unique. Once it has been published and is in the public domain, it should not be reproduced – even in a very slightly altered form – anywhere else except under exceptional circum-stances, for example being included in a celebratory issue marking some dignitary's retirement. Even so, whoever holds the copyright should give permission. As the editor of an international journal, I had a case where a paper was brought to my attention by another editor that had an uncanny resemblance to a manuscript that I had published a year earlier. Indeed, almost everything was the same, five figures having been slightly massaged so that the values were not exactly identical. But there was one difference that made this second article (published in a different journal) not entirely a 'duplicate' in that it dealt with a different cell line from the first! It was also surprising that the second paper came from a different institution from the first, although both were in mainland China. The first paper provided an exact template for the second, which meant that very minimal changes were needed to deal with a different cell type that appeared to have given these very similar data. This could suggest fabrication as well as duplication, and at the least it clearly suggests sheer laziness on the part of the authors.

Some authors have a habit of repeating many of the particulars from a previous paper before extending them with a little new information; this is effectively 'self-plagiarism' (see the next sections). This is unwarranted and amounts to duplication where it is extensive. If the same editor sees the second paper, this would lead to rejection unless it could be improved to include essentially only the new data. However, the second paper might have gone to a different journal and the editor would probably be unaware of the extent of duplication; it should then be more easily spotted at peer review. The possibility is that, if 'iThenticate' software (the www.ithenticate .com/) is not used by the editors, peer reviewers and/or publishers, the paper could slip through the net and be published. One thing is for sure; this will have wasted the editor's and publisher's time, and done nothing to advance the status of the author(s). The conclusion is that duplication, fully or partially, is unethical and should not be practised at any cost. The next subsection deals with another aspect of this matter.

16.2.2 Multiple Submissions

A paper should be submitted to only one journal at a time. The authors have to wait until it has been rejected by the first before they can legitimately submit to another journal. Modern software increasingly recognizes multiple submissions, so be warned. Referring to both multiple submissions and duplication, there are some practices that can less easily be detected; let us take the case of a paper sent to an international journal to be published in English, and at or about the same time it is submitted to a national journal, let us say to be published in Japanese in Japan, and therefore only for a more limited readership. This is unethical, but it still occurs too often. If your paper gets published in an international journal in this day and age, it is hardly worth publishing it nationally, and therefore there is no point in being unethical. A paper written in Japanese is unlikely to be cited by readers worldwide when they write papers on the subject; there is no gain in this kind of duplication/multiple submission.

16.2.3 Plagiarism

Plagiarism is theft from previously published papers of text, figures and tables, indeed any part of other people's intellectual property that is seen as published articles. This is an infringement of copyright. It even includes acts of self-plagiarism, as mentioned in Section 16.2.1. It also refers to ideas stolen from others, but this is more difficult in many cases to monitor. These practices are highly unethical and should be avoided at all cost. There is in fact no excuse for it, because permission can usually be granted by authors and publishing houses who hold the copyright, as long as acknowledgement is made and references given to any part of an article you might wish to include in your own paper. This procedure is almost always free of charge, at least in the scientific literature dealing with research papers and reviews. Publishers have small departments dealing with permissions, but can impose restrictions on the length of reproduction of another person's work. Thus taking material in this way ought to be kept to the very minimum.

Modern editing software is now sufficiently sophisticated to detect most of the plagiarism that occurs. The iThenticate software is commonly used, as has already been mentioned, especially when there is any suspicion that plagiarism is occurring in a new submission, and sometimes a simple browser search will pick up problems.

16.2.4 Fabrication

Fabrication of data is truly a major offence and can lead to rejection of all future works by an author. In some flagrant cases it has led to criminal proceedings and even imprisonment (the Hwang Affair in South Korea, in which nine papers, each dealing with different aspects of stem cells, were mostly fabricated (see https://bit.ly/3egKACv)). The authors tried to divert suspicion in this case by publishing their papers over a period of about a year in more than eight different journals.

The authorities at the institutions of authors are informed of malpractice and asked to reprimand their staff who committed the fabrication (as is the case with duplication and multiple submissions). If the situation is bad and appropriate action is taken, then the guilty person(s) may well be dismissed and their careers effectively halted. Unfortunately it is not often known whether disciplinary action has been taken, especially where the level of fabrication is not extensive. The reason this ought to be more universally enforced is that, unless fabrication reaches criminal proportions, as in the Hwang case, international bodies monitoring malpractice usually have neither the authority nor the means to take litigious action in most countries.

Extreme cases that have led to prosecution are few and far between, but that does not mean that fabrication on a smaller scale is uncommon. Authors can be under pressure to publish, and there may be small experiments or perhaps bits of data needed to help support a shaky hypothesis. To complete a report quickly, some such 'facts' might be concocted to fill the gaps. There is no easy way of checking fabrication at this level; editors and publishers have to rely initially on the integrity and honesty of authors and the opinion of peer reviewers. Fabrication may include taking reasonable guesses at what might have been found in reality, but if incorrect, later papers could rectify the problem and no one would necessarily be any the wiser. The better policy is not to indulge in it in any way in the first place. But there is another form of fabrication, also dishonest and seemingly more trivial, that is covered in the next section.

16.2.5 Manipulation of Data and Information

It is sometimes inviting to make the most of a data set that might be illustrated, for example, as a curve in a graph. However, there could be 20 points that all seem to show excellent correlation with another parameter, but 2 that seem totally unruly. There is a natural instinct to omit these 'rogue' points, but refer back to Chapter 9 to recall why this practice is scientifically unreasonable; a reader wants to see *all* the experimental data, and can sometimes make more of them than the author.

There are literally hundreds of ways in which data and information can be manipulated in reporting research work. This can occur by leaving out data or adding inadmissible points, values, data or even ideas. Images can be airbrushed and computer-manipulated to massage them into a form that helps make some point or other; this almost amounts to fabrication. But there are equally cases where image manipulation does help to make some data more distinct. Let us take an obvious example just to illustrate this point. Whenever we see posters of high magnification images of microbes or intracellular structures taken by scanning or transmission electron microscopy, they are not in black and white as they were originally, but are often artificially coloured for effect. This is not seen as misrepresentation, but a coloured electron micrograph in a scientific paper would not be condoned. Perhaps one of the most common transgressions is to alter the appearance of something where a strict comparison is needed based on the original data, e.g. the density of a band in a Western blot may be changed, or it could be cut and pasted from another with the comparison with neighbouring bands being from the same separation. Unacceptable airbrushing can often be detected and should be avoided. If the experimental work has been done professionally and the results properly presented, there should be no need for manipulation. To be fully truthful, if the data you are presenting are worth publishing but do not seem to fit comfortably together, you ought to point out things like rogue points and other findings that seem to weaken any correlation and your hypotheses; they can then be included in the Discussion.

There is always some give and take in findings, especially in biology, which is why statistics are needed (more often than in the physical sciences), to show the probability of correlation rather than ever expecting to have a perfect correlation. The bottom line here is do not mess with the data; publish what you have found and be prepared to explain it when it does not seem that tight. Better still, repeat the work until you have firmer data; that is why researchers are expected to do no fewer than three runs on each part of an experiment. If this does not get you any further, tackling the problem from a different perspective may yield good data. This leads us on to negative or neutral findings and their significance in preparing a paper for submission.

16.2.6 The Problems of Negative and Neutral Findings

We have just dealt with the removal of some rogue points from a graph as being unethical. Extending this, we are probably all guilty of trying to prove the hypothesis under examination, and this usually means selecting the data from experiments that help support it. This process quite naturally tends to make us guilty of ignoring negative and neutral data (this criticism can probably be levelled at all of us, as you will see later). Data selection is where some experiment may be repeated, say, four or five times and only three give consistent results of a significant difference between the treated and control groups, whereas the other two were neutral, i.e. showed no effect. Are these two ignored, just brushed aside? Sometimes things go quite 'wrong' on such occasions, the results either being neutral or suggesting a negative rather than a positive correlation. This makes things even more troublesome, but now it is more difficult to claim that there is indeed any correlation at all. Indeed, in both cases these findings should alert the experimenter to the fact that the hypothesis itself does not hold any water and needs to be revised or rejected. People often remark on 'that one niggling little fact' found during a set of experiments that simply does not fit in with the rest of the experimental series in support of a hypothesis. If the pressure is not too great to publish, then the best plan by far is to hold back from submitting; clearly the work is not yet ready for publication.

In this last discussion, we would all be behaving differently in science and medicine (as in other fields) if we approach things according to the teachings of the philosopher Karl Popper (https://en.wikipedia.org/wiki/Karl_Popper), who repeatedly pointed out that we should be trying *to disprove* our hypotheses, not going all out to prove them. If a hypothesis stands after repeated attempts to disprove it, there must be some truth in it. How many of the papers published today go down this route? Only on a few occasions does clear *disproof* of a hypothesis get published, and that is usually when it has been part of the received wisdom and is long-standing. It is when the whole field of knowledge has been upset and has to be reconsidered from a different perspective; think of Galileo and the geocentric idea that had prevailed before, regarding the rotation of the Sun about the Earth. (Turning a hypothesis upside down can be a sensible exercise to do at any time that one is investigating some phenomenon!) The issue is that it is not only authors who might deal with negative results inappropriately, but editors and publishers who seem generally to be more biased towards accepting papers with positive results, and often have no time for those with either neutral or negative findings. The bias introduced is a topic that is seldom discussed and no clear policy has been formulated to correct it. To give an example of this bias, let us take the case of clinical trials. If only trials that yielded positive results (for example, that a particular drug was effective against a certain disease) were reported, when as many (if not more) showed no efficacy at all, they most certainly should not be ignored. Attempts are now being made to get all trials reported in clinical circumstances to allow a balanced informed conclusion to emerge. Would that this practice could now be extended to all aspects of experimental work and its reporting!

16.3   Other Places to Consider Regarding Unethical Practices

16.3.1   Acknowledgements

Although customs have changed since it used to be courteous to acknowledge everyone from the most junior technician to the secretary and typist involved in the production of a paper, today we expect authors to acknowledge anyone who has contributed material and materially to the body or main thrust of the paper. We ought to include people who have offered cultures, specimens, samples, etc. to the project, those experts who have provided insights that were highly relevant and people who had contributed significantly to the scientific and/or medical content of the article.

The biggest transgression is not to acknowledge all the sources that made the project possible. These include government agencies, charities, private donations, venture capitalists and so on. Specific details are needed, such as names of government agencies and grant numbers. It is highly discourteous and morally negligent to omit the very means that allowed the article to be moved into the public domain.

16.3.2   Retractions and Withdrawals

There are undoubtedly issues relating to the fickleness of authors regarding the destiny of a potential publication. If, after submission, continued work shows its purport to be possibly misconstrued, incorrect or blatantly wrong, there is little other option than to request *withdrawal* before the paper gets to the acceptance stage. To let it go ahead such that the authors can set the record straight in later communications might be a way out, but it is morally indefensible. Withdrawal is a better option, although the matter must be firmly agreed with editors and publishers before it is too late. If the paper has gone past acceptance or even been published, then authors can ask for a *retraction*. Incorrect data in a paper, rather than some more urgent reason to remove the whole article, can be dealt with by submitting corrections (corrigenda), which are often published in a subsequent issue of the same journal. This indicates that the authors have acknowledged a major fault in the data, incorrectness in its presentation in the published version, or inconsistency in the logic of the argument(s) to the point that they would rather not want it taken as having thorough integrity as a meaningful piece of work. To let a paper go on without making a retraction or providing corrigenda is therefore quite unethical.

16.3.3   Permissions (Including Clinical Data)

One clearly improper procedure is to fail to acknowledge permission for the publication of any material quoted or paraphrased from an existing article. While this has been covered in detail under copyright (Chapter 15), there are other circumstances in which permissions are all important, and if neglected can lead to big moral and ethical issues that can land authors up in court. One very sensitive one is where enough information is given about the name, status or description of a person discussed in a clinical case study that he or she could be recognized by others – loss of anonymity. No one wants this to happen, and it must be avoided at all cost.

Similarly, questionnaires used extensively in healthcare and similar studies report on the behaviour of different groups of people. These must always be discussed with potential participants in a study beforehand. There must be informed consent, a document which preserves the anonymity of the participant and allows the researchers to proceed without fear of litigation. It is always better to think ahead of time whether what you are doing, and what you want to achieve, can go ahead without any redress from all those involved in a study or project. Planning is all important, and in some cases so sensitive that it is wise to involve the help of people who are widely experienced in the field of study that might otherwise throw up a plethora of problems. The prospect of litigation is always best avoided by being scrupulous about the intentions and procedures of your studies ahead of time; it is too late to consider the possible consequences after having set out.

16.3.4   Co-Authorship Issues

Co-authors are often seen as minor players in the bigger drama; some even have their names tagged on without contributing anything more than being the head of a unit. This might seem scurrilous; but it happens, sometimes because of deference to higher authorities, in respect to one's elders, the mores of some cultures, and so on. Authorship, in terms of those deemed to have been properly involved in the work, i.e. contributed materially and philosophically to the article, must be carefully considered before submission (refer back to Chapter 11). All co-authors need to be fully informed, know about the submitted article in detail and have given their consent to its publication. There are many instances in which a submitted or published paper has been challenged by one of the authors, a head of department, a principal investigator or someone in a similar position. Take the case of a postgraduate student with a good article to publish from his or her PhD work, that is subsequently published in his or her name without reference to others who contributed materially to it in the place where the work had previously been done. The supervisor might become quite incensed and matters not proceed well, depending on how far such a paper has progressed before the problem confronts the editor of the journal and/or the publisher; trying to resolve these problems is far from easy and is a burden that ought to have been dealt with before submission ever was contemplated. There is no easy resolution in these cases, and the unethical behaviour of the author moves out of the hands of the editor and publisher, who have no option other than to stop any further action until the conflict has been resolved. Their easiest action is simply to reject the manuscript.

Another problem is where co-authorship has not been acknowledged, i.e. someone has contributed to the work but not been included. While some corrections are possible, many papers will have got past the point of introducing any change, although some mention can be made in a later issue of the same journal, as a corrigendum, mentioned in Section 16.3.2, although this is not seen as giving due respect to the omitted author.

16.3.5 Conflicts of Interest

You might consider the above discussion as being a conflict of interest, but here we more often are referring to situations in which an author has some allegiance to a body, an organization or a person that compromises their independence with regard to the substance of a paper. If I have a vested interest in BioProng Standards Ltd which has funded a research project in which I have been involved, the published information might be greatly to my own benefit and that of the company. Not to declare an interest is immoral. The ethics of publishing must clearly be based on disclosure of any vested interest in the work. We are dealing here with scientific and medical literature (as is much the same elsewhere), which has to be 'squeaky-clean', or in blunt terms not just for profit.

16.3.6   Commercial Pressure: Predatory Open-Access Journals and Publishers

Perhaps one of the inevitable consequences of online publishing is that it allows an editor or publisher to put through papers of weaker scientific or medical integrity on the 'author pays' model because commercially it is a highly viable and profitable exercise. Many nefarious organizations have taken advantage of this situation. *Predatory* journals try to seduce authors with cheaper online publication of their data using poor (or false/fake) reviewing methods, trying to 'milk the market'. Thirty years ago it was thought that many established publishing houses 'subsidized' their continuing production of learned journals from profits made from popular novels, etc. This scenario changed soon after, and scientific and medical publishing is now big business and highly competitive. On the 'author pays' model, the publisher can assure you that you will get your article published after acceptance once you have paid the article processing charge (APC). There is a greater incentive to let weak papers through as it simply means more revenue. The moral and ethical dilemma is that publishing greed compromises scientific integrity. Let us leave this conundrum by considering the philosophy of Vitek Trasz, founder of BioMed Central in London, who considered that we might just as well publish anything and everything that authors submit *and are willing to pay to have published* (the author-pays model), and leave it to the readers to decide what is good and what is trash. This was from a businessman who knew little about science and was concerned with commercial gain. If this policy is widely adopted, the whole of scientific integrity and worthiness is put in jeopardy; this comes at a price – integrity must be protected at all cost, but certainly not for commercial gain. There is much more at stake that needs to be guarded and treasured, and herein lies the whole basis on which the ethics of scientific publication stands. Predatory journal editors invite potential authors to submit – but it is important not to be seduced.

16.4   Help Is At Hand

There are organizations that will help in solving ethical problems that might arise. Most are voluntary bodies that are there to assist authors and others in coping with problems that might lead to ethical issues. Coping is the mot juste, since one of these organizations is called **COPE** (Committee on Publication Ethics). Other similar bodies are there to 'police' the activity of editors and publishers such that they also respect the problems of authors, including **EASE** – the European Association of Scientific Editors, **WAME** – the World Association of Medical Editors, **CSE** – the Council of Science Editors, and others. Thus there is no reason for an author to confront problems that cannot be surmounted in an ethical manner. If you break the rules, you will have to endure now and perhaps for many years hence the penalties of being unethical in your approach to scientific publishing. It is better to be ethical than indulge in any form of malpractice. Advances in science require maximum integrity; if this is lacking, it is not only you but the whole of science that suffers.

17 Epilogue

A research article is *the final product* of an investigation. It is a report that tells the world what you have done and found. On this account, it must be presented in the best possible way so that its message is to the point, clear and succinct.

I should have craved your indulgence by now for using the same banner at many of the earlier chapter headings, as here. However, it is there to be a constant reminder to you that committing your work to publication is the only sustainable way in which people will learn about it and how it advances our scientific and general knowledge. Its importance should never be underestimated, which is why we must be properly trained in this art of communicating our research work.

17.1 Cardinal Points

Let me apologize to those of you who have noticed the repetitive nature of some of the information in the many previous chapters. *It was not unintentional because many readers will consult this book chapter by chapter rather than reading it through at a single sitting.* The idea was to make each chapter as standalone as possible so the reader doesn't have to keep jumping from one chapter (or section) to another. To complete this book, or more appropriately 'manual', on writing good scientific papers, I condense here what may be seen as the 12 cardinal points. If these are kept in mind, the task of preparing a paper becomes easier. Although Chapter 1 provided an overview, it did not emphasize these essential points, neither was this done in the following chapters as each dealt with only its smaller compass.

The 12 cardinal points are:

(1) Always have something original to write.

(2) Come straight to the point.

(3) Be as succinct as possible at all times and do not overload your paper in either data or words.

(4) Your readers are usually experts in the same field; don't be banal or condescending in your Introduction or elsewhere.

(5) Ensure that the evidence for (or against) your hypothesis being tested – the point around which everything revolves – is scientifically and logically sound.

(6) Write the paper in the following order: Results, Discussion, Introduction, Materials and Methods, Abstract, Title. The other sections can then be slotted in, i.e. References, Acknowledgements, Figure Legends and Tables, with the figures themselves being appended as separate files, as also any Supplementary Information.

(7) Ensure that all your co-authors have read and contributed to the paper, including making corrections; you will also need to acknowledge any other help you were given, including the funding body.

(8) Check whether you are about to submit to an 'author pays' or 'reader pays' journal.

(9) If the English is weak and ought to be improved to increase your chances of acceptance, find an expert in your field who is a native English speaker to check it out, or send it to a professional editing service that deals in improving manuscripts before submission.

(10) Submit only after complying implicitly with the journal's Instructions to Authors, preferably after using a checklist.

(11) Respond favourably and responsibly to editors, reviewers and publishers; they are trying to help you and disputes are counterproductive.

(12) Never submit your paper to more than one journal at a time, just as it is unethical to duplicate papers, plagiarize others and fabricate data.

At this point, it is also worth revisiting page 15 to see the flow chart covering the main points set out above.

17.2   Last Words

There are far too few courses training people to write scientific papers, which is a tragedy because this is the product of any research which requires much skill to accomplish satisfactorily. Most institutions pay little more than lip-service to this issue, even the best only giving perhaps one or two short lectures on the topic. Most people going on to research will inevitably have to learn the art, but most seem to have learned by osmosis. Many copy what others have done in their published works, which is alright up to a point, but many bad habits arise because you are often reading papers written by others that are poor in their presentation and content. Doing so means that you will never properly develop a style of your own, which becomes obvious after reading a plethora of already published papers. It is also best to seek advice and attend any courses that can help, wherever they may be held.

It takes time to write a good scientific paper, especially in high quality idiomatic English. Indeed, it is usually more difficult to write a top-class short paper than a long one. This is because the first draft will need to go through many revisions and be honed until it is succinct and lucid. This pleases editors, reviewers and readers enormously, which is greatly to your advantage. So take your time; it is better not to rush into submission! A really good succinct paper is read by reviewers and editors much more quickly, making its chances of being accepted far higher than a paper that is poorly presented.

Performance will also improve with practice and this is by far the best way to gain the experience that makes writing papers often a pleasure rather than a chore. Since a satisfactory career in research depends on your products – your published papers – the importance of writing good papers cannot be overemphasized.

I have written specifically about the preparation of the standard primary research article. More could be said about writing short communications and preliminary reports, but these should essentially be done in the same manner. However, I have resisted dealing with (i) technical reports; (ii) abstracts for meetings; (iii) review articles; (iv) theses; and (v) articles for consumption by the general public. In further editions of this manual, I hope that some information that helps in their preparation can be included. Perhaps the most difficult is to 'translate' the average scientific paper to the general public, a job which is often left to journalists, but which is seldom attempted by scientists. To communicate the message accurately and not in a distorted manner takes exceptional skill, and many articles fail because they fall between two stools, with certain inadequacies of insight in scientific journalism and the reluctance, as well as inadequacy, of most scientists to make a good job of these communications.

To assist you, some of my essays on the presentation of papers in English are also available for download. In another shorter book, I will deal specifically with this art of using words effectively (see under 'Writing Style' in the Further Reading section for introductory help). But this will not be just for scientific presentations. It will help non-native English speakers and those who have English as their second language. For some, it is advisable to get help first by going to English courses run in many cities and towns around the world before attempting to put scientific and medical papers into idiomatic English, as this is a more difficult task that involves much jargon. It is this, units of measurement, acronyms and many other contrivances in the language used in research communications that make it extremely difficult for the layperson to make sense of even the simplest scientific paper. If an author of such a paper can keep this in mind at all times, communication becomes much more effective at all levels to all people.

Further Reading

As mentioned at the beginning of the manual, literature references have been purposely left out of the text to avoid confusion and inconsistencies. However, it is worth consulting articles by other authors on many of the topics to gain a wider perspective of the problems. Many books and articles can be found online about writing scientific papers, and also on the procedures of submission and publication. Few of them are as comprehensive as this manual in dealing with all three aspects but they offer useful additional advice.

The listings below refer you to sources that are particularly relevant and useful. Four short lists will cover:

(1) Preparing a paper.
(2) Procedural matters in submission and publishing, including ethical considerations.
(3) Advice on how to write good idiomatic English and develop style.
(4) Essays and comments by the author that can be helpful extensions to the topics in the main text.

This manual has not dealt with issues arising in (3), especially where authors are non-native English speakers. This would require another manual of much the same size, and there are already many that deal specifically with this issue.

Preparing a Paper

Burrough-Boenisch J. (2013). Editing texts by non-native speakers of English, in Smart P, Maisonneuve H and Polderman A (eds.) *Science Editors' Handbook*, 2nd Edition. European Association of Science Editors. Available online at www.ease.org.uk/publications/science-editors-hand book. (This gives the editorial perspective that can also help authors.)

Diskin S. (2018). *The 21st Century Guide to Writing Articles in Biomedical Sciences.* Singapore: World Scientific.

EASE (European Association of Science Editors). (2018). Guidelines for authors and translators of scientific articles to be published in English. *European Science Editing,* **44**(4), e2–e16. Although this journal is for editors and does not patently serve authors, the guidelines are now published online (www.ease.org.uk/publications/author-guidelines) in 29 languages. Its structure follows much of the layout of our manual, but in a potted version notably over the first six pages, thereafter being a series of tips on smaller matters (e.g. ethics checklist, unusual plurals in English).

Gastel B and Day R. (2016). *How to Write and Publish a Scientific Paper,* 8th Edition. Santa Barbara: ABC-CLIO.

Hoffmann AH and Isaac AO. (2014). *Writing in the Biological Sciences,* 1st Edition. Oxford: Oxford University Press.

Kotyk A. (1999). *Quantities, Symbols, Units and Abbreviations: a Guide for Authors and Editors.* Totowa, NJ: Humana Press.

Matthews JR. (2014). *Successful Scientific Writing: A Step-By-Step Guide for the Biological and Medical Sciences,* 4th Edition. Cambridge: Cambridge University Press.

Weissberg R and Buker S. (1990). *Writing Up Research.* Englewood Cliffs, NJ: Prentice Hall.

Zeiger M. (1999). *Essentials of Writing Biomedical Research Papers,* 2nd Edition. New York: McGraw-Hill.

Submitting and Publishing a Paper

Chipperfield L, Citrome L, Clark J et al. (2010). Authors' submission toolkit: a practical guide to getting your research published. *Current Medical Research & Opinion,* **26**(8), 1967–1982, DOI: 10.1185/03007995.2010.499344.

Committee on Publication Ethics (COPE). (2020). Promoting integrity in scholarly research and its publication. Website of COPE available at https://publicationethics.org.

Council of Science Editors. (2018). *Scientific Style and Format: the CSE Manual for Authors, Editors and Publishers,* 8th Edition. Chicago, IL: University of Chicago Press. Information and subscription options available online at www.scientificstyleandformat.org/Home.html.

Michaelson H. (1990). *How to Write and Publish Engineering Papers and Reports*. Phoenix, AZ: Oryx Press.

Nature Research. (2020). Image integrity and standards. Website available at www.nature.com/nature-research/editorial-policies/image-integrity.

Writing Style

Cutts M. (2013). *Oxford Guide to Plain English*, 4th Edition. Oxford: Oxford University Press.

Glasman-Deal H. (2010). *Science Research Writing: for Non-Native Speakers of English*. London: Imperial College Press.

Gowers R and Gowers E. (2015). *Plain Words*. London: Penguin Books. (N.B. The first author is continuing to produce new editions. This book was written many decades ago because, as head of the British Civil Service, Ernest got fed up with jargon, verbose and highfalutin words and phrases in memoranda being sent him. He asked that staff write in simple and succinct words; to be straightforward and to the point – no 'bureaucratese'. The same applies to scientific reporting of results.)

Kozak M and Hartley J. (2019). Academic writing: an inconsiderate genre. *European Science Editing*, **45**(3), 69–71.

Marsh D. (2013). *For Who the Bell Tolls*, 1st Edition. London: Guardian Books.

Essays on Individual Topics by the Author

The author also suggests the reader looks at his **essays and comments** on several topics:

Wheatley D. (2012). On the current presentation of scientific papers: 1. Editing out redundancy. *European Science Editing*, **38**(4), 97.

Wheatley D. (2013). On the current presentation of scientific papers: 2. Cutting out clichés. *European Science Editing*, **39**(1), 13.

Wheatley D. (2013). On the current presentation of scientific papers: 3. Referencing. *European Science Editing*, **39**(2), 41.

Wheatley D. (2013). On the current presentation of scientific papers: 4. Spacing things out. *European Science Editing*, **39**(3), 70.

Wheatley D. (2013). On the current presentation of scientific papers: 5. Verbs and tenses. *European Science Editing*, **39**(4), 99.

Wheatley D. (2014). Plagiarism: a prevalent and persistent problem. *European Science Editing*, **40**(4), 69–70.

Wheatley D. (2014). English as the *Lingua Franca* of science. *European Science Editing*, **40**(2), 40.

Wheatley D. (2014). Drama in research papers. *European Science Editing*, **40** (1), 14.

Wheatley D. (2016). Why aren't researchers taught how to write and publish papers? *The Biologist*, **63**(5), 7.

Wheatley D. (2018). Writing scientific and medical papers clearly. *The Anatomical Record*, **301**, 1493–1496.

Index